PUBLIC HEALTH ENGINEERING

SEWERAGE

Second Edition

PUBLIC HEALTH ENGINEERING

SEWERAGE

Second Edition

RONALD E. BARTLETT
F.I.C.E., F.I.P.H.E., F.I.W.E.S., M.Inst.W.P.C.
Consulting Civil Engineer

APPLIED SCIENCE PUBLISHERS LTD
LONDON

APPLIED SCIENCE PUBLISHERS LTD
RIPPLE ROAD, BARKING, ESSEX, ENGLAND

British Library Cataloguing in Publication Data

Bartlett, Ronald Ernest
Public health engineering, sewerage.—2nd ed.
1. Sewage disposal—Great Britain 2. Sewage —Purification
I. Title II. Sewerage
628′.3′0941 TD557

ISBN 0-85334-796-4

WITH 37 ILLUSTRATIONS, 40 TABLES and 9 CHARTS

Originally published in 1970 as
Public Health Engineering—Design in Metric—Sewerage

Printed in Great Britain by Galliard (Printers) Ltd, Great Yarmouth

Preface to the Second Edition

SINCE the publication of the first edition in 1970, the metric (SI) system has become fully established in this country and in many countries overseas as the system of units in use in the civil engineering industry. Amendments have been made in the text to take account of this, and the original chapter relating to the changeover to metric units has been deleted.

A new chapter has been included to deal with the more specialized aspects of house drainage and smaller sewerage schemes, and the opportunity has been taken to correct a few comparatively minor errors of the earlier edition.

Since 1970 there have been considerable changes in the standards of materials used in civil engineering. A number of British Standard Specifications have been withdrawn and new Standards have been issued. New types of pipe materials have come into use and the earlier standards for iron and asbestos-cement pipes have been revised.

Public health engineering (as with all branches of civil engineering) is continually developing, and this continuous process must, of necessity, be reflected in a new edition of a book of this nature. In that respect, the author has made a number of changes to the section devoted to the early preparation of schemes, as there have been many new developments following the re-organisation of local government in the UK.

ASHBY-DE-LA-ZOUCH
LEICESTERSHIRE
1978

R.E.B.

Contents

List of Formulae

List of Tables

List of Figures

1 Introduction

WORKS of sanitation are known to have existed in ancient times, and archaeologists have shown that drainage systems were in use in Roman cities over 2000 years ago. Despite this, there was really no effort to provide sanitation in the United Kingdom and other modern countries until the nineteenth century. The systems of sewerage as we now understand them have developed from experience gained only over the last 100 years or so.

Sewerage has really developed since the advent of piped water supply and the consequent need for an effective means of removal of waste waters. The design and construction of sewerage works is one part of the work of the public health engineer.

THE PUBLIC HEALTH ENGINEER

The public health engineer is usually also a qualified civil or municipal engineer. The Institution of Public Health Engineers (originally founded as the Institute of Sanitary Engineers in 1895) serves as a centre for the promotion and distribution of knowledge of public health engineering. That Institution describes public health engineering as 'a positive force in the field of preventative medicine': 'By providing a supply of pure water, conveying and disposing of sewage and refuse, removing storm water so as to prevent flooding, and by raising the standard of environmental hygiene by improving plumbing, drainage, heating and ventilation, public health engineers have been and continue to be instrumental in bringing about great advances in living standards.'

The Institution defines the profession as 'The art and science of designing, supervising, executing and administering work intending to assist, develop and control the forces of nature in order to maintain and improve the health of the community'. It also defines the public health engineer as 'A person who is engaged as an engineer in connection with public health and who designs, or controls, or undertakes, or advises upon constructional works or other like matters affecting the health of the community'.

Public health engineering is generally understood to include the design and construction of works of water supply, sewerage, sewage treatment and disposal, and refuse disposal. Design in all of these facets of engineering involves many variables, which make public health engineering much more empirical than many other branches of engineering. This branch of civil engineering is also one which cannot be practised in isolation; it probably entails more teamwork than most other branches, as co-operation is usually necessary with structural, electrical and mechanical engineers, architects, town planners, geologists and meteorologists; in addition, very close liaison is maintained with the medical profession, and with those engaged in chemistry, biology and bacteriology.

SEWERAGE

Sewerage is defined in BS CP 2005 as 'a system of sewers and ancillary works to convey sewage from its point of origin to a treatment works or other place of disposal'. A sewerage system is one of a number of vital public utilities upon which the modern community is so dependent. Unfortunately, so much of this work is not visible to the general public, and the absence of glamour often results in a lack of appreciation of its importance.

Sewers (together with all the 'ancillary' works, including pumping installations) are designed to collect and convey both domestic and industrial waterborne wastes, and surface water run-off. Foul sewerage design is based on the number and density of buildings, the number of families per building, the size of the family, and the varying habits of the population as regards the use of water.

For the adequate design of surface water sewers, the engineer must have a knowledge of the topography, together with details of the intensities of rainfall of the particular district. There is an economic limit to the intensity of storm that can be catered for, and in practice sewers are usually designed for the worst storm likely to occur every year or, in some cases, every five years. Various design methods for surface water sewers have been used in the past, but modern practice is either to use the 'rational' (Lloyd-Davies) formula or the TRRL hydrograph method (see Road Note 35) [15].

In view of the flows involved, a surface water sewerage system is usually designed as a gravity system, pumping being kept to the minimum. For a foul sewerage scheme, it is often necessary to compare the alternative engineering considerations and economics of gravity sewers and a pumping scheme. Pumping may, in fact, be the only practicable solution in a flat district.

The hydraulic design of sewers—the calculation of capacities and velocities of flow—has become well established in recent years. The basis of the formulae and tables prepared towards the end of the last century has been confirmed by experience and by more recent research.

The structural design of buried pipelines was developed in America early this century and has been in use in the UK for a number of years. The importance of structural design became more apparent with the development of flexible/mechanical joints for all types of pipes and with the increasing use of plastics in engineering. Most local authorities now require hydraulic *and* structural calculations for any new system of sewers.

CONTRACTS AND CONSTRUCTION

The issue of revised Codes of Practice and the preparation of a number of Working Party Reports have drawn attention to the many modern developments in sewerage design and construction. The importance of an adequate specification and of good site supervision to supplement these modern methods has been stressed.

While the legal aspect of a contract must be maintained, the design and construction of a sewerage scheme is becoming more and more a matter of co-operation and co-ordination between designer, manufacturer and contractor. This can only be to the benefit of the client, as he *should* then obtain a better job at a lower price.

LEGISLATION

During recent years the approach to drainage and sewerage design has been affected by new legislation. The more important recent Acts of Parliament include the Public Health Act, 1961, and the Rivers (Prevention and Pollution) Acts, 1951 and 1961. The latest Public Health Act abolished local building by-laws and empowered the Minister to make regulations in their place; new regulations were published in 1965 and these have been republished as the Building Regulations, 1976.

The Public Health Act, 1961 brought up-to-date the legislation on the discharge of trade effluents. It allows local authorities to specify conditions attached to any consent to discharge trade waste to a sewer, and to make a charge for its reception and disposal.

The Rivers (Prevention of Pollution) Acts established a system of control over new discharges of trade and sewage effluents to streams and rivers, and the Water Act, 1973 made these subject to the consent of the Water Authority. The conditions which attach to new discharges as regards quality, temperature, volume, etc., can also be applied to earlier discharges.

2 Site Investigations

INVESTIGATIONS and surveys form the basis of assessment of a site for the works envisaged, and they lead to the engineering decisions upon which the ultimate design is based. Surveys and investigations must therefore be carried out competently and thoroughly if an adequate and economic design is to be achieved.

In generally accepted terms, 'investigations' include the collection of information on such matters as access, other services, ownerships, geology, climatic conditions, etc., while the term 'survey' usually refers to the use of precise instruments for the measurement of positions and levels. To some extent, and in the present context, the two expressions are more or less synonymous.

Initial surveys and investigations will be of a broad nature, so that various possible outline schemes can be compared. Later surveys to form the basis of a specific scheme will be more precise and detailed.

Surveys for a sewerage project should be undertaken by engineers experienced in sewerage design, as their knowledge of the various aspects of this type of work will enable them to recognize the information required. This information will be obtained from existing maps and aerial photographs, town-planning proposals, records of existing works, borings and trial holes, and instrument surveys, together with detailed personal examinations of the development, industries, climate and contours of the area to be sewered.

PRELIMINARY RECONNAISSANCE

A preliminary reconnaissance will determine the natural drainage areas in the district of the proposed scheme, and will provide a general indication of the geological conditions.

This initial reconnaissance will determine whether existing properties have 'combined' or 'separate' drainage systems (see Chapter 7) and whether further development is likely in the foreseeable future. It will also determine the type of development (residential, commercial or industrial) in the area, and the general types of industrial wastes to be expected.

The approximate flows in streams and rivers should be ascertained so that they can be considered with a view to using them to take the discharges from any proposed treatment works or storm sewage overflows. Approximate information on flood levels can be noted from the banks themselves. Any special uses of the rivers for water supply intakes, ornamental waters, etc., must be recorded. Where relevant, the effect of tidal flows is important.

Early consideration must be given to the possibility of more than one point of discharge or treatment for the sewage from a large area, or, alternatively, the feasibility of a joint scheme which could cater for more than one local authority. Local authority boundaries will hardly ever coincide with the watersheds of drainage catchments, and the Water Authority may have to agree the extent of any new proposals, to avoid the duplication of sewers and treatment works. On a smaller scale, the same reasoning can be applied to the drainage of housing estates, factories or farms, where co-operation with neighbouring establishments may often show a saving in either capital or running costs.

Although it may be possible to do much of the early reconnaissance from a car by driving along adjacent roads and tracks, it will often be necessary to walk over much of the ground, so that such features as quarries and cuttings can be examined in detail. This inspection will also provide an opportunity for noting any obstructions, such as pipelines and overhead cables, which are not marked on the available maps, and also for plotting such features as road diversions and new buildings which have been constructed since the Ordnance Survey maps were last revised. General information on the apparent positions of other services should be noted (for confirmation later) together with details of any roads with high traffic densities.

Should it be considered necessary to carry out any levelling or other instrument survey work at this stage, it is generally wise to establish frequent survey points and temporary bench-marks throughout the area under consideration. These reference points can then be used as the basis for any later detailed surveys, without the necessity for further extensive 'tying in' to known points.

The aim of this preliminary reconnaissance must be to obtain sufficient information to supplement existing maps and records, so that outline proposals can be drafted for any alternative solutions. Comparative estimates of capital and annual costs then can be made, as these will usually form the basis of any ultimate choice of scheme.

MAPS AND PLANS

The Ordnance Survey maps of Great Britain contain considerable information which can form the basis for planning a sewerage scheme. In some overseas countries the available maps may provide much less information; if so, time must be allowed in the initial period for more extensive site investigations and instrument surveys, together with aerial photography where relevant. In 1978, the Department of Transport issued a 'Model Contract Document for Topographical Survey Contracts'; the Specification section includes detailed requirements for permanent bench-marks and aerial surveys, and for conventional signs to be used on plans.

Ordnance Survey maps are usually available at scales of 1:50 000 and 1:100 000 (or the nearest equivalents) for the preliminary investigations of the routes of sewers, while maps at the larger scales of 1:10 000 and 1:25 000 can be used for all field reconnaissance work. For the final survey plans, Ordnance Survey sheets at 1:2500 or 1:2000 should be used. Sections along the proposed sewers will then eventually be plotted, using either 1:2500 or 1:2000 for the horizontal scale.

Schemes to be submitted to the Water Authority for approval should usually be accompanied by key plans to a scale of 1:10 000, although 1:25 000 will be acceptable if the area covered is large. Drawings of proposed structures, such as pumping stations, sewers, special manholes, etc., should be in sufficient detail to enable the purpose of the works and the form of their construction to be readily ascertained. All proposals for land purchase should be shown on plans to a scale of 1:2500.

Geological Survey maps are at present published in Great Britain to scales of 1:50 000 and 1:10 000, together with descriptive memoirs, reports and handbooks. Some older maps to 1:63 360 and 1:10 560 are still available. These maps and supporting documents, along with records of the National Coal Board (where relevant), should provide an adequate picture of the underground strata for the general planning of sewer lines and the location of treatment works. They can be supplemented by information obtained later from borings and trial pits.

Where works are to be carried out in the sea or in harbours (e.g. sea outfall pipes) reference should be made to the relevant Admiralty hydrographic charts and to the local tide tables.

CLIMATE

The design of a sewerage scheme is affected by rainfall intensities and by temperatures. To some extent, the 'prevailing wind' will also affect the siting of pumping stations and treatment works.

In Great Britain, while the amounts of total annual rainfall vary from nearly 5000 mm (in North Wales and the English Lake District) to under 500 mm (along parts of the East coast), the intensities of storms throughout the country can be expected to be roughly similar. The design of a surface water sewerage scheme is based on the intensities of storms of short duration; the actual length of storm to be considered will depend on the size and type of the drainage area, and may vary from a few minutes to a few hours.

It is general to use one of the published formulae which give probable rainfall intensities in terms of duration of the storm. In some circumstances these formulae may not be suitable, particularly for schemes outside Great Britain, and it will then be necessary to collect and analyse records of storms, so that the rainfall intensities for various relevant storm durations can be calculated (see Chapter 6).

Temperature statistics will affect the choice of materials for pipelines and other works, and also the minimum gradients to be adopted. In Great Btitain the usually accepted basis of sewer design takes into account the normal temperature range, but for overseas schemes it may be necessary to consider any possible effect of local temperatures.

High temperatures will result in an increase in the generation of hydrogen sulphide in the sewers, and therefore an increase in the possibility of corrosion of certain materials. When very low temperatures are to be expected, or where the air temperature can go below 0 °C for any prolonged period, this may affect the construction timetable, the depth of cover to be given to pipelines, and the installation and maintenance of machinery.

POPULATION STATISTICS

Before any assessment can be made of probable flows of foul sewage, the designer must have a reasonably accurate assessment of the present and probable future populations. These will affect the calculations of flow in a 'separate' system of sewers, and also the degree of dilution to be expected in a 'combined' system.

Estimates of existing population figures can be built up from information obtained during the last available census and by reference to the registers of voters. These registers, of course, only include the names of persons entitled to vote. The information obtained in this way can be supplemented from the records maintained by the local authority of new properties which have been built in the district.

Estimates of future populations in developing areas will depend to some extent on the natural increase in the population of the country as a whole (the increase of births over deaths). Future population estimates in Great Britain are usually based on existing figures, with an allowance (either up or down) which will take account of any town-planning proposals for the area. These estimates can then be adjusted to provide for any particular local tendencies for development.

Any sewerage system should be designed for at least the population to be expected twenty years hence. In view of the time which usually elapses between the initial design and its final commissioning, it is often customary to formulate sewerage proposals on the estimated population of at least thirty years ahead. It may, of course, be very difficult to estimate this with any accuracy, but it should be borne in mind that any loan sanction will be based on a period of forty years for the sewers.

EXISTING SEWERS AND WATERCOURSES

While some local authorities have very full records of existing sewers in their areas, showing their diameters, positions, levels, etc., there will be many instances where this information is not available. There are many older sewers for which no records are available, and it is often found that records of private sewers are either non-existent or are very inaccurate.

Copies of all available records should be collected very early during the initial investigations, and where necessary these must be checked and supplemented by detailed inspections on site. As many old sewers will be devoid of manholes, or the covers will be hidden or difficult to open, this type of inspection can be a very lengthy and tedious task.

Before any calculations can be prepared for a new scheme, the designer should have full details of the location and levels of the existing system. This should include information on sewer sizes, and whether they are 'separate' or 'combined'. He must also have details of existing flows and the capacities of the existing sewers, together with any information on their liability to surcharge or to collect silt.

Where there are existing pumping stations or ejector stations, details of their capacities must be obtained, together with a note of any spare capacity, and any provision of space in the station for extensions to the installation. Full information must be obtained on the diameter, location and capacity of all rising mains. and the condition of the sewage after pumping (e.g. whether septic).

From the financial viewpoint, it will be necessary to consider the effects of any outstanding loans on sewers, plant or buildings, and the running costs of existing installations.

Any investigation into an existing system of sewers must, of necessity, also embrace an investigation of the existing watercourses in the catchment area. The ditches, streams and rivers which form the natural drainage of the district will set the pattern for a surface water sewerage system, and will also provide sites for treatment works or storm sewage overflows where these are to be included.

Sufficient information should be collected on site to enable a reasonable estimate to be made of the capacity of any watercourse, and of the 'normal' and 'flood' flows. Physical, chemical and bacteriological examinations should be made of the flows when relevant. Methods of sampling for these types of examination are described in 'Notes on Water Pollution' No. 8 [21].

GEOLOGY

During any site investigation, the soils and rocks encountered must be classified by an accepted, concise and reasonably systematic method, so that the information collected can be used correctly, so that useful conclusions can be drawn from it and so that comparisons can be made with other sites or projects.

The Geological Survey maps referred to earlier give a good general guide to the types of soils or rocks to be expected on any site. Information on the geological history of the area (e.g. whether volcanic, alluvial plain or reclaimed land) will also be very useful. This must then be supplemented by borings or trial pits, depending on the importance of the proposed structure.

The names and references given to the soils and rocks encountered in civil engineering investigations are not necessarily related to the nomenclature adopted in other scientific terminologies. Appendix C of BSCP 2001, 'Site Investigations', contains a list of definitions of soils and rocks which has been drawn up from the standpoint of the engineer. Appendix D of the same Code sets out information of special interest to engineers on geological processes, modes of formation and structures.

Much information can be gleaned from records of earlier excavations in the area and from personal inspections of quarries, cuttings, etc. This early collection of existing information based on a 'geological reconnaissance' will usually provide a suitable background for the initial feasibility studies and estimates. Trial holes and/or borings can then form part of a more detailed site investigation before the final designs are completed and before the bills of quantities are prepared.

TRIAL HOLES AND BORINGS

Once the proposed layout of sewers has been more or less decided, a detailed survey of the nature of the subsoil along their proposed routes is possible. This information is needed for the structural design of the pipeline (see Chapter 10) and so that proper provision can be made during the contract for trench support and for dewatering. The usual form of conditions of contract (see Chapter 3) assumes that the designer has provided the tenderer with information on subsoil conditions and his rates in the tender will be based to some extent on that information..

The importance of subsoil investigations for any scheme cannot be overstressed. The engineer responsible for the design must have clear authority for determining the extent, etc., of these investigations; the full findings should be made available to all tenderers.

A number of firms have specialized in site investigation work, particularly in the sinking of boreholes. These firms will quote for drilling through various strata (at rates per metre depth), with lump-sum rates for the erection and dismantling of the rig, taking undisturbed samples, and carrying out various tests on the excavated soil, either on site or at their own laboratories. These borehole surveys can be very useful, particularly for plotting the underground contours of the various strata.

Site investigations are, however, often undertaken after submitting a competitive tender, and the contractor may be more interested in borehole depth at the fastest speed possible, resulting in a sacrifice of accuracy in the description of subsoils. If water is used to assist the drilling, it is also possible to obtain misleading information on subsoil water levels.

For sewerage work, and particularly where pumping stations and sewage treatment works are included, the information obtained by boreholes should normally be supplemented by a reasonable number of trial holes. These will indicate the type of side support which will be needed at excavations, and will give a clearer picture of subsoil water conditions. In clay soils, deep explorations may best be carried out from borings and shallow explorations from pits. In sandy soils, boring is easy, but special equipment will be needed if undisturbed samples are required.

When trial pits are excavated, these can sometimes be left open for a long period so that a study can be made of the effects on the subsoil water level during periods of increased run-off in watercourses or during variations of the tide. The proximity of subsoil water level to the surface may affect the design of the sewers, as in some circumstances it may be more economical to lay shallow sewers (with pumping stations) to avoid the expense of deep excavations. On the other hand, in some types of clay subsoil it may be possible to carry out comparatively deep excavations to (say) 4 or 5 m below the water level in adjacent watercourses without experiencing any great difficulty with subsoil water, and with only light supports to the sides of excavations.

SURVEY EQUIPMENT

The general term 'surveying' covers a wide field and includes aerial surveys and land surveying with theodolite and tacheometer. Surveying for a sewerage scheme will normally only entail the use of an engineer's level, along with a chain or tape, but more complete surveys will be necessary (including aerial surveys) when large-scale maps are not available.

The newer engineer's levels have enclosed 'split image' bubbles and can be supplied with or without a horizontal circle. Over recent years, the standards of accuracy of this type of equipment have improved, eliminating the human element to a great extent. Self-aligning levels are now available in which the collimating line of sight is no longer levelled by hand (with a spirit level), but in which it is levelled automatically by pendulum prisms, within a limited range of instrument tilt. The choice of instrument will depend on the type of work for which it is intended. A small builder's type of dumpy level is of little use for sewer survey work, and the instrument chosen should preferably have a range of up to 150 to 200 m (for estimation of readings to an accuracy of 1·0 mm). Such an instrument would have an accuracy of $\pm$ 2·0 mm per kilometre.

Levelling staffs are available in many patterns, and the traditional telescopic staff has now lost its popularity in favour of the lighter folding staff, particularly the type which will close up to a shorter length for transport. A 4-m telescopic staff weighs about 4 kg and is 1·5 m long when closed. To allow for the length of staff which may be below ground when taking levels of manhole inverts, a 4-m staff should be used for sewerage surveys. The lighter types of 4-m staff can be obtained to close to 1·0 or 1·2 m, a length easily accommodated in the boot of a car. Metal staffs are to be preferred where they are likely to be used in manholes or other wet conditions. Staffs are normally graduated for direct reading to 10 mm, allowing estimation down to 1·0 mm.

While an engineer's chain will prove useful for an extended survey over rough ground, much use is made of measuring tapes for most sewerage work. The linen tape has lost its earlier popularity, and the choice is usually between a steel tape or one of fibreglass. Where the extreme accuracy of a steel tape is not warranted, a fibreglass tape is usually used; it is more or less unaffected by wet conditions and is easily cleaned after use. A steel tape should be used for all important measurements where accuracy is a vital factor, e.g. setting out and laying foundations and other constructional work. Chains are manufactured in 20-m and 30-m lengths, while most tapes are obtainable in lengths of 10, 15, 20, 25 or 30 m (some manufacturers produce 40- and 50-m tapes). Equipment for most sewerage surveys should include a 30-m chain, together with a 30-m tape.

LEVELLING

The object of 'levelling' is to establish the elevations of different points, or to establish the difference in elevation between a number of points. If the actual elevations (related to Ordnance datum) are

Date 1st January 1970 Levels taken for Newtown devel. – sewers
From High St. To West St. PATTERN A

BACK SIGHT	INTER-MEDIATE	FORE SIGHT	RISE	FALL	REDUCED LEVEL		DISTANCE	REMARKS
2.97					105.92			O.B.M. on school
1.40		0.76	2.21		108.13			C.P.
	1.81			0.41	107.72			Cover – Ex. M.H No. 33
	3.93			2.12	105.60			I.L. — " —
	2.15		1.78		107.38			I.L. 150 mm ϕ culvert by church
3.80		0.35	1.80		109.18			C.P.
	2.51		1.29		110.47			℄ junc. exist. estate road
	2.02		0.49		110.96			I.L. last M.H on — " —
	0.57		1.45		112.41			Cover — " — — " —
	1.21			0.64	111.77			I.L. Exist. factory drain
3.51		0.17	1.04		112.81			C.P.
	2.80		0.71		113.52			G.L. corner of site 'A'
	3.71			0.91	112.61			I.L. ditch at corner
	2.51		1.20		113.81			T.B.M. (see sketch)
0.28		3.51		1.00	112.81			C.P. (previous)
0.42		3.61		3.33	109.48			C.P.
1.08		2.55		2.13	107.35			C.P.
		2.53		1.45	105.90			OBM (105.92)
13.46		13.48	11.97	11.99				
	(0.02)		(0.02)		(0.02)			

Fig. 2.1. Booking levels—rise and fall method.

required, then work must start from a point of known elevation—this may be an Ordnance Survey bench-mark or a temporary bench-mark. The temporary bench-mark (TBM) may have been established during an earlier survey, or it can be related to some known level, such as a factory floor, to which has been allocated a value above an assumed datum. This adoption of an assumed datum is convenient on a self-contained site, where the design of the drainage is not affected by the levels of other work off the site.

Levels are booked in level books on either the 'rise and fall' method or on the 'line of collimation' method. The latter method should be used when the values of levels at change points are more significant than those of other points, but this method should *never* be used for sewerage work. When taking levels for sewers, change points are often incidental (having been established to enable the main survey to make progress) and the values of intermediate points may often be more important. The 'rise and fall' method should be used for all sewerage work.

Although the method of booking levels cannot obviously affect the accuracy of the survey work, the rise and fall method of booking checks the arithmetic of *each* entry, so that any error in reducing a level will show up at the end of the survey. This is not so with the line of collimation method. As a check on the accuracy of reduced levels, the four columns for 'backsight', 'foresight', 'rise' and 'fall' should be totalled for each page. The difference between the totals of backsights and foresights should equal the difference between the totals of rises and falls, if the arithmetic is correct. This difference will also equal the difference between the first and last reduced levels on the page. This method of booking levels is illustrated in Fig. 2.1.

When levelling for sewerage work, the aim should be to have a closing error (in millimetres) of not more than $20\sqrt{L}$, where L is the length of the line of levels in kilometres. For example, if a line of levels from a bench-mark back to the same bench-mark is a total of 4·0 km, then the final closing error should not be greater than $20\sqrt{4} = 40$ mm, or 0·04 m. This is considered too conservative by some engineers, who accept closing errors of from 40 to 100 mm per kilometre.

3 The Design Office

THE preparation for, design and construction of any civil engineering project in the United Kingdom usually follows a well-defined pattern, whether the promoter is a government body, a local authority or private enterprise.

After the initial proposal, an engineer's report is prepared and approval must be given to the estimated expenditure. Sewerage schemes prepared by a local authority will be subject to the approval of the Water Authority and will form a part of that Authority's Capital Works Programme. It is usual to prepare the estimates, reports, drawings, etc., with that in mind.

There are very many stages in the planning and organization of a contract for a sewerage scheme and ample time must be allowed for these when preparing details of a new scheme.

The amount of work to be carried out in the design office will depend to some extent on the type of scheme. A design which will be submitted to the Water Authority for approval and will then ultimately go out to competitive tender will usually require more drawings and a more detailed specification than one which will be carried out as part of a package deal within a commercial organization.

ECONOMICS

A sewerage or drainage scheme is not directly revenue-producing, but the local authority or company for whom the design is being carried out will need to have accurate estimates of both capital and running costs. This is particularly important where the capital cost must be borrowed, as provision must be made for the repayment of both capital and interest.

Throughout the whole of the design process, from initial survey to the final submission of a completed scheme, the engineer must bear in mind both the technical and financial implications of possible alternatives. On a sewerage scheme, this will entail the examination of alternative routes for sewers, the choice of suitable materials, and the pros and cons of different diameters and gradients. The choice may be between open cut and tunnelling or between deep gravity sewers and pumping.

It will be seen that the design engineer must keep up-to-date on costs of materials and construction so that alternatives can be sensibly compared, from both technical and cost standpoints. An understanding of the fundamental mechanism of finance is also necessary if the engineer is to apply his technical experience with economic advantage. From the latter point of view, the designer is recommended to study a booklet entitled *An Introduction to Engineering Economics*, published by the Institution of Civil Engineers [35].

A proper understanding of the economic issues on a scheme will ensure that the engineer provides adequate information on which a tender can be based. It will also enable him to analyse tenders. This is particularly important for pumping and similar installations, where tenderers will be providing guarantees of performance and will be submitting tenders based on their own designs.

SUBMISSION OF A SCHEME TO A WATER AUTHORITY

As the details of many sewerage schemes are prepared for ultimate submission to a Water Authority, it is usual for all calculations, estimates, reports and drawings for local authority schemes to be prepared with that in mind.

Schemes will be 'vetted' as to their administrative, financial and technical soundness and, depending on the existence of any restrictions on expenditure, it may be necessary to put forward a case on the urgency of the need for the scheme. The information to be submitted will include a technical report supported by a key plan—these should both be clear and concise.

At any subsequent local investigation, the authority should present a complete, concise, well-organized case, with all available supporting evidence. In a paper to the Association of Rural District Council Surveyors in 1968 [43] Best said:

> A good presentation would describe the existing facilities and underline their defects; it would emphasize and explain the need for and likely effect of the proposed improvement and indicate the urgency; it would describe the proposals clearly and explain why they are thought to be the most suitable of the alternatives available; it would indicate the general financial implications of the project; and it would confirm that the necessary supporting consents and other permissions required could be got on acceptable terms. The presentation would also include a site demonstration of the suitability of the proposals.

Technical details should include information on the size and type of pipes, particulars of joints, bases of calculations of flow to be carried, capacities of sewers proposed, any formula or graph used for calculating surface water run-off, drainage area plan, number and siting of storm sewage overflows. Where pumping stations are included, information is required on static and frictional heads on pumps, pump duties and velocities in rising mains.

This information can either be included in the engineer's report, on the standard form or as a separate statement. Where relevant, further information is required on the basis of the design of reinforced concrete structures and on the location of trial holes and the information recorded.

If land acquisition is envisaged in the scheme, this must generally be settled before approval can be given, so that any action by the Authority will not prejudice the consideration of a compulsory purchase order.

MEASUREMENT

The Institution of Civil Engineers' 'Standard Method of Measurement' specifies the units of measurement to be used in civil engineering bills of quantities and the basis of division within those units for the purpose of measurement.

There are three units in common usage which are not included among the basic SI units. These are the litre (l), the tonne (t), and the hectare (ha). Use of these three units avoids the unnecessary use of prefixes. The 'litre' and 'hectare' have been used throughout this book, along with the 'hour' and the 'cumec' where these are relevant.

All dimensions on drawings and in 'taking off' should be in the preferred units, and standards of accuracy must be related to the work being detailed or measured. On drawings, dimensions should normally be entirely in metres (to three places of decimals where relevant) or entirely in millimetres. Where the units used are obvious, and where the above 'three places of decimals' rule has been applied, there should be no necessity for the inclusion of any reference to the units, but if there could be any doubt the unit of measurement should be stated.

It has been suggested that as a general rule each measurement should normally be taken to the nearest 10 mm (i.e. 5 mm and over to be regarded as 10 mm, and less than 5 mm to be disregarded). There must, of course, be many exceptions to that rule, particularly concerning accuracy of setting out and construction, as distinct from measurement for payment. It will normally be necessary to set holding-down bolts and similar items to the nearest 1 or 2 mm, while finished levels of concrete, etc., may be required to the nearest 5 mm, or perhaps 2 mm. Depths to invert of pipes (for assessing excavation rates) could be taken to the nearest 100 mm, while lengths may be taken to the nearest metre.

On all documents, the decimal position should be marked by a decimal point (·). A thousands marker should *not* be used, and figures on either side of the decimal point should be grouped in sets of three with a single space between each set of digits (e.g. 98 082·156 24). BS 1957, 'The Presentation of Numerical Values', permits the use of four digits together.

ORDNANCE SURVEY

Ordnance Survey maps form the basis of most preliminary layouts for sewerage schemes. Normally, use is made of the maps to 1:1250 and 1:2500 scales for local details, and 1:10 000 for the overall layout of schemes. Occasionally the 1:25 000 and 1:50 000 maps will be useful for planning regional schemes.

The National Grid Reference System is now used on all modern Ordnance Survey maps; the basic grid is of 100 km squares, while larger-scale maps are divided into 1-km squares (1:50 000), and 100-m squares (1:2500 and 1:1250). Values of bench-marks and contours are given in metres, but in view of the instability of some bench-marks (e.g. due to subsidence), the Ordnance Survey may in future omit the values of bench-marks from large-scale maps. Up-to-date bench-mark lists are available (on repayment) from the Ordnance Survey.

Areas of parcels of land on the 1:2500-scale maps are frequently given in both hectares (metric) and acres (imperial). When these figures are used for the estimation of areas for run-off calculations using the formulae quoted in this book, care should be taken to use the metric figure. Existing maps, with areas quoted in acres, will no doubt be in use for another twenty years or more, and use can then be made of the conversion factor set out in Appendix F.

Enlargements to 1:500 scale of 1:1250 maps, and to 1:1250 scale of 1:2500 maps can be supplied by the Ordnance Survey. These enlargements are printed on 0·18-mm-thick sheets of polyester plastic approximately 1000 mm by 600 mm.

LAYOUT OF DRAWINGS

Drawings related to a sewerage scheme are usually prepared to form part of a set of contract documents. They may also be produced to illustrate reports or as final 'as constructed' records. Whatever their purpose, it is usual to standardize on sizes and scales and on the general layout of the details of the drawings, so that the information is presented in a neat, orderly form which is readily understood.

The majority of drawings prepared for sewerage schemes in the past have been on 'antiquarian', 'double elephant' or 'imperial' size paper. The approximate sizes of those papers, and of corresponding drawing boards, are given in Table 3.1, along with sizes of drawing sheets in the International Standards Organization 'A' Series.

TABLE 3.1
SIZES OF DRAWING PAPERS

Designation	*Size (mm)*	
	Drawing paper	*Drawing board*
Antiquarian	1 350 × 790	1 370 × 810
Double elephant	1 020 × 670	1 070 × 740
Imperial	760 × 560	810 × 580
ISO A0	1 189 × 841	
ISO A1	841 × 594	
ISO A2	594 × 420	
ISO A3	420 × 297	
ISO A4	297 × 210	

The older type horizontal plan chests are generally based on the storage of double elephant or antiquarian drawings. Vertical plan-filing systems are available to take drawings to the ISO 'A' series. While it is probable that many offices engaged on sewerage design may eventually adopt size A1 as their standard drawing size, it would seem prudent to use filing cabinets large enough to hold the A0 size. The cabinets would then conveniently house copies of older drawings or drawings received from other offices; or two A1 drawings could be filed side by side on the same holder.

When drawings are to form a part of a set of documents to be submitted to another authority for approval, all drawings and plans (except Ordnance maps) should be on strong paper and should be folded separately to go into a standard 325 × 230 mm (C4) envelope and should be numbered. Drawings must be fully dimensioned with scales and, where applicable, north points clearly indicated. Longitudinal sections should run in the same direction as the plans and should be plotted to the same horizontal scale. It will generally be found preferable to have the drawing title, scale, number and other relevant information in a panel not more than 200 mm long at either the top or the bottom right-hand corner of the drawing, so that drawings can then be folded 'concertina' fashion, with this panel visible.

Sections along sewers should show existing ground levels, proposed invert levels, gradients, pipe diameters and other details in a clear form, and this should be standardized throughout all drawings. A suggested layout is given in Fig. 3.1.

DRAWING SCALES

The preferred scales are reproduced at Table 3.2. For sewerage schemes, it will be rare to use the scales referred to under the headings 'Assembly' and 'Details', although these might be used by

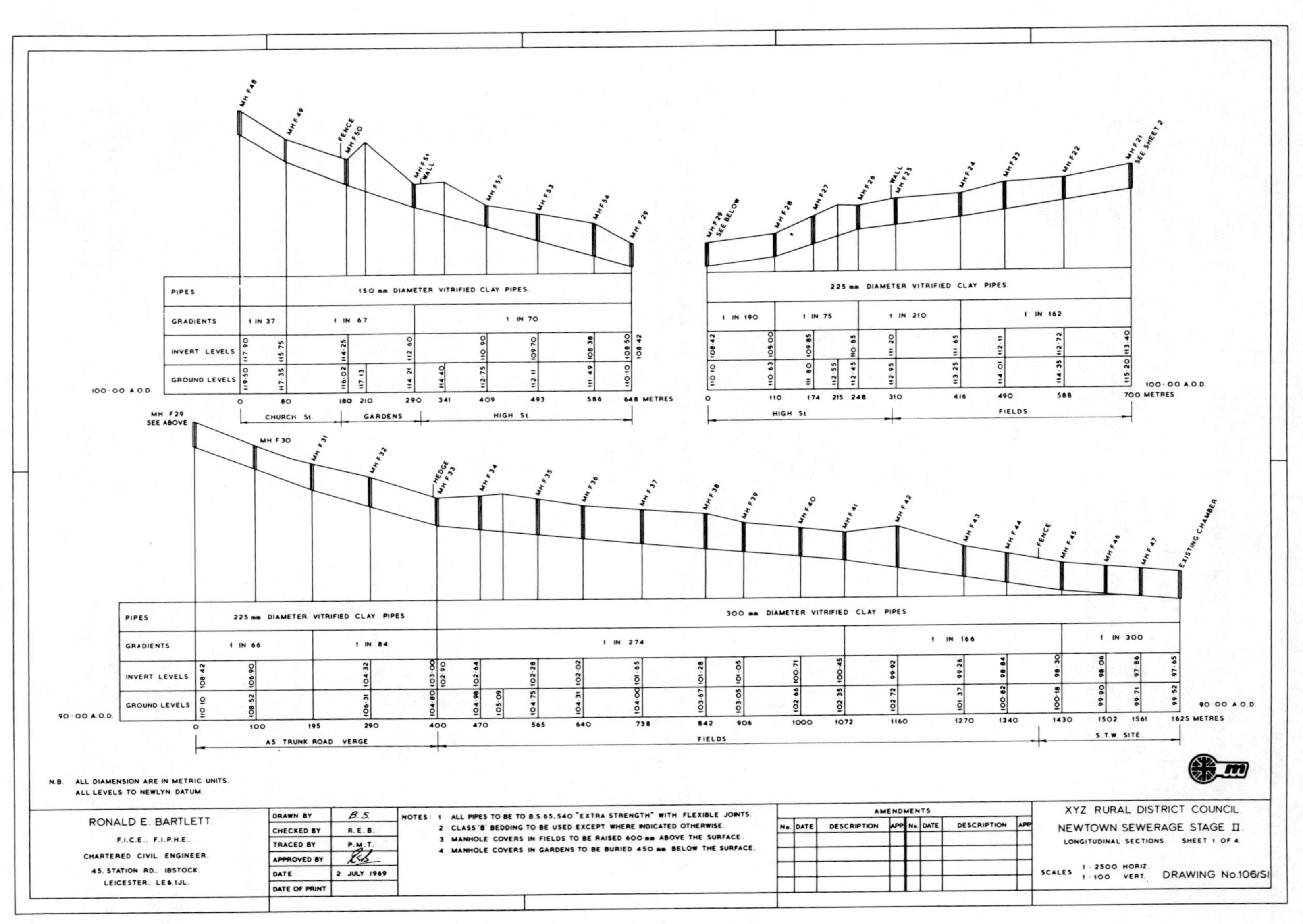

Fig. 3.1. *Typical sewer sections.*

TABLE 3.2
PREFERRED SCALES

Use		*Scale*
Maps		1:1 000 000 1:500 000 1:200 000 1:100 000 1:50 000
Town surveys		1:50 000 1:20 000 1:10 000 1:5 000 1:2 500 1:2 000
Location drawings	Block plan	1:2 500 1:2 000 1:1 250 1:1 000
	Site plan	1:500 1:200
	General location	1:200 1:100 1:50
Component drawings	Ranges	1:100 1:50 1:20
	Assembly	1:20 1:10 1:5
	Details	1:10 1:5 1:1

This extract from PD 6031 is reproduced by permission of the BSI, 2 Park Street, London, W1Y 4AA, from whom copies of the complete document may be obtained.

manufacturers of equipment. The scales under the heading 'Ranges' will be used for building details, drawings of concrete work, manholes, vertical scales on sewer sections, etc.

Key plans to illustrate lines of sewers and the general layout of a scheme should be to a scale of 1:10 000, but if the area covered is large a scale of 1:20 000 would be acceptable. Plans and sections of sewers will normally be drawn to 1:2500 (vertical scale of sections probably at 1:100), with manhole details at 1:20. Scales of drawings of pumping station details will depend on the size of the building, but might conveniently be to 1:50, with details to 1:10.

During the years when it is possible for drawings to exist in both metric and imperial units in the same office, it is essential that all drawings are clearly marked with the system of units used.

Recommended marking methods are included in BS 308, although the use of the metric 'key' symbol will assist in rapid identification of metric drawings. Samples of the key symbol can be obtained from the Press and Public Relations Department, BSI, 2 Park Street, London, W.1. Its use is indicated in Fig. 3.1.

With all 'ratio' scales, scaling from drawings is possible with one simple boxwood scale with the edge divided into millimetres, as the conversion to actual dimensions is direct, e.g. a length of 25 mm on a drawing to a scale of 1:20 is equivalent to 25 × 20 = 500 mm on site. In some circumstances it will be simpler to use specially prepared scales which will read off in metres from drawings to such scales as 1:1250 and 1:2500.

CONTRACT DOCUMENTS

The several documents which together form the contract include the conditions of contract, the specification, the bills of quantities and the drawings, along with the tender, the agreement and the bond. The first two of these are considered in more detail in Chapters 4 and 5.

Drawings have been referred to earlier in this chapter. Sufficient drawings are normally prepared initially to form a basis for a contract, and to enable quantities to be taken off for incorporation in the bills of quantities. Further drawings will usually be prepared by the engineer during the course of the contract.

The conditions of contract (Clause 1) state that the term 'drawings' means 'the drawings referred to in the specification and any modification of such drawings approved in writing by the Engineer, and such other drawings as may from time to time be furnished or approved in writing by the Engineer'. It is therefore important that a list of drawings should be prepared and issued with the specification.

The bills of quantities will usually be divided into a set of 'preliminary' items, followed by one 'bill' for each section of the works. The method used to divide up the work into bills is a matter of personal preference to some extent, but on a sewerage scheme these will usually fall into the following groups:

1. Sewers—one or more bills.
2. Manholes.
3. Ancillary works—storm overflows, etc.
4. Pumping stations.
5. Rising mains.
6. House connections.
7. Sewage treatment works—subdivided into units as required.

The 'preliminaries' will usually include separate items for each of the clauses in the conditions of contract, together with any other general items, such as resident engineer's office (including heating and attendance), watching and lighting, traffic control, temporary roads, holidays with pay, insurance, payment of rates, etc. These items are included to enable the contractor to price them separately if allowance has not been made in his unit rates. It is usually specifically stated that if no sum is entered against a particular item it is understood that the cost of that work or service has been included in the unit rates.

The specification and bills of quantities are frequently bound together as one document. This document will also contain any instructions to tenderers and the 'conditions of tender', the list of drawings (if not actually printed in the specification itself), and the form of tender. The conditions of contract, form of agreement and form of bond are normally those published by the Institution of Civil Engineers, and it is usual to refer to these without actually reproducing them.

Where selective tendering is adopted (see Chapter 12) the additional expense of a performance bond may not be necessary. While each contract must be considered on its merits, the bond should be dispensed with in all cases wherever this seems reasonable.

4 Materials

THE value of an up-to-date specification was referred to in Chapter 2. It is prepared to enable a contractor to assess the standards of materials and workmanship required by the engineer, so that he can price his tender. The specification is also a legal document and must not therefore be open to misinterpretation. Where relevant, it should state quite specifically that materials or workmanship '*shall be*' to a certain standard; expressions such as 'to be' should not appear in the document. Where the employer or the engineer is to provide a service, the words 'will be' or 'may be' can be used.

The conditions of contract, the specification, the bills of quantities and the drawings form a set of contract documents, and care must be taken to ensure that they agree. The specification should cover all items billed and may also cover other materials or work not billed if there is reason to believe that these might be required during the contract. Details shown on the drawings must agree with the specification and the bills of quantities.

BRITISH STANDARD SPECIFICATIONS

Many of the materials to be incorporated into a sewerage contract are fully described in BS specifications. In such cases, it is not usual to describe fully the material in the specification, but only to refer to the relevant part of the British Standard.

When work is to be in accordance with a BS code of practice, all materials and components should comply with the latest edition of a British Standard where applicable. Clause B2 of the Building Regulations, 1976, states that the use of any material or method which conforms with a British Standard or a BS code of practice 'shall be deemed to be a sufficient compliance with the requirements' of the Regulations. All pipes, manhole covers, gullies and other drain fittings must comply with the current requirements of the relevant British Standard when used on housing schemes covered by the guarantee of the National House-building Council.

Care is required in quoting British Standards, however, and where one standard covers a number of qualities or sizes, the quality or type required must be sufficiently described in the specification. It is general either to refer to 'the latest BS' in all cases (i.e. the latest at the time of tendering) or to quote the actual date of issue of each British Standard. It is bad practice to mix the two methods in one specification, as this can lead to confusion during the contract.

A list of British Standards relevant to sewerage work is included at Appendix B. The reader should refer to BSCP 2005 for a more detailed list.

PIPES—GENERALLY

With modern methods of pipeline design and construction, it is recommended that only flexible or flexibly jointed pipes should be used for both sewers and rising mains. Rigid-jointed pipes are still manufactured in vitrified clay, concrete and cast iron, but their use is very limited. Rigid-jointed pipes should *not* be used in conjunction with a granular bedding or in ground which is subject to settlement. The rubber rings used with flexibly jointed pipes should comply with BS 2494.

The choice of pipe material will depend to some extent on personal preference. Vitrified clay pipes are suitable for most conditions. Neither vitrified clay nor concrete pipes should be used in the close vicinity of strongly rooted trees, or as suspended pipelines in basements, etc. Concrete pipes are subject to attack from sulphates under certain conditions (see below and also Chapter 11). Pitch-fibre pipes should not be used where they will be subjected to hot effluents nor at shallow depths.

All pipes should be handled and stacked on site in accordance with the manufacturer's instructions, and care must be taken to avoid damage to factory-applied joints. Any rubber rings should be stored carefully to avoid damage and so that there will be no confusion between various manufacturers or between different diameters. As far as possible, vitrified clay and concrete pipes should not be stacked on site where they will be subjected to uneven temperatures which could cause cracking.

VITRIFIED CLAY PIPES

Vitrified clay drain and sewer pipes have been manufactured since about 1845 and are the most widely used for gravity sewers. Standards of manufacture are covered by BS 65 and 540, while BS 539 gives dimensions of fittings. Some manufacturers can supply pipes with additional chemically resistant properties (BS 1143).

BS 65 and 540 include two types of pipe, viz. 'British Standard' quality for general use and 'British Standard Surface Water' quality for surface water sewers only. Both types of pipe are available as either 'standard strength' or 'extra strength' and both types can be supplied with flexible joints.

Vitrified clay pipes have safe crushing-test strengths as set out in Table 4.1, and their approximate weights (in terms of aggregate length of pipes per tonne) are given in Table 4.2. In addition to the diameters listed in Table 4.1, they can be supplied in diameters of 125, 175, 200, 375, 400, 450, 500 and 600 mm at special request. They are available in various lengths up to about 1·5 m and are now not normally glazed.

The crushing strength tests provided for in BS 65 and 540, 556 and 3656 are all based on the use of a test machine with rigid top and bottom bearers fitted with rubber bearing strips. The method of

TABLE 4.1
SAFE CRUSHING TEST STRENGTH OF VITRIFIED CLAY PIPES TO BS 65 AND 540

Nominal diameter mm	*Test strength per metre of inside length*					
	Standard strength		*Extra strength*		*Super strength*	
	kgf/m	N/m	kgf/m	N/m	kgf/m	N/m
100	2 000	19 600	2 200	21 600	2 800	27 500
150	2 000	19 600	2 200	21 600	2 800	27 500
225	2 000	19 600	2 800	27 500	—	—
300	2 200	21 600	3 400	33 400	—	—

TABLE 4.2
WEIGHTS OF VITRIFIED CLAY PIPES

Nominal diameter mm	*Approximate length of pipes metres per tonne*	
	Standard strength	*Extra strength*
100	76	70
150	45	42
225	27	25
300	15	14
375	9	7
450	6	—

manufacture of vitrified clay pipes does not permit the production of absolutely straight pipes and there is a growing feeling that for this type of pipe the testing machine bearers should be flexible, so that they can assume the same profile as the pipe itself. Research has been carried out in Sweden using a hydraulic hose positioned between a series of nine movable segments and the loading and supporting beams. In the USA and in the UK a programme of research has used a conventional crushing-test machine, equipped with compensating segmented loading and supporting devices.

Vitrified clay pipes up to 450-mm diameter can be obtained with factory-applied 'push-fit' joints. These incorporate polyester farings and a rubber 'O' ring. As an alternative, on smaller diameters, flexible joints can be obtained using plain-ended pipes with plastic push-fit sleeve couplings. Pipes can also still be obtained with plain sockets for use with rigid sand/cement joints. For land drainage work, perforated vitrified clay pipes are available to BS 65 and 540; these are manufactured to 'extra strength' quality with perforations in one half of the circumference only, and in plain-ended lengths up to 1·5 m.

Pipes with chemically resistant properties manufactured to BS 1143 generally conform to the requirements of BS 65 and 540, in terms of dimensions and tolerances. The hydraulic test in BS 1143 is, however, more severe, while additional tests are included for absorption and acid resistance.

CONCRETE PIPES

Concrete pipes are manufactured to BS 556 and can be either unreinforced or reinforced, according to the strength classification. They are available in internal diameters from 150 to 900 mm in 75-mm increments, and from 900 to 3000 mm in 150-mm increments. The minimum proof and ultimate crushing-test loads for the four new classes of pipe are given in Table 4.3. Pipes of nominal diameter larger than 1800 mm are available for any agreed strength, and pipes of higher minimum crushing strength are also available for any standard nominal diameter.

Pipes can be supplied in various lengths, according to manufacturer and diameter, from approximately 1 to 2·5 metres. A schedule of approximate weights of pipes with flexible joints is given in Table 4.4.

TABLE 4.3
MINIMUM CRUSHING TEST LOADS OF CONCRETE PIPES TO BS 556 (LOADS IN kgf/m)

Nominal diameter mm	Standard		Class 'L'		Class 'M'		Class 'H'	
	Proof	Ult.	Proof	Ult.	Proof	Ult.	Proof	Ult.
150	2010	2380	—	—	—	—	—	—
225	2010	2380	—	—	—	—	—	—
300	2010	2380	—	—	—	—	2380	2980
375	2010	2380	—	—	3120	3900	3720	4650
450	2010	2380	—	—	3570	4460	4170	5210
525	2010	2380	—	—	3870	4840	4610	5760
600	2010	2380	—	—	4610	5760	5510	6870
675	2010	2380	—	—	5060	6320	6100	7620
750	2010	2380	3870	4840	5360	6700	6550	8190
825	2010	2380	4170	5210	5810	7250	6990	8740
900	2010	2380	4610	5760	6850	8560	8630	10790
1050	2010	2380	5210	6500	7740	9670	9820	12280
1200	2010	2380	5800	7250	8780	10980	11160	13940
1350	2010	2380	6400	8000	9680	12100	12360	15430
1500	2010	2380	6990	8780	10560	13200	13400	16750
1650	2010	2380	7580	9480	11750	14690	14880	18600
1800	2010	2380	8330	10420	12650	15800	16070	20090
1875	2010	2380	8780	10980	13100	16370	16670	20830
1950	2010	2380	9080	11350	13540	16940	17110	21400
2025	2010	2380	—	—	—	—	—	—

By courtesy of the Concrete Pipe Association.

In addition to spigot and socket pipes with various proprietary forms of flexible joint, concrete pipes are also made with ogee joints to BS 4101. The BS includes a crushing-proof test load of 2010 kgf/m of length for straight pipes for one minute 'without sign of failure' and a hydraulic proof test of 4000 kgf/m^2 for half a minute 'without leakage, but appearance of moisture permitted'. Ogee joints are not watertight, and these pipes are therefore suitable for use as culverts and where infiltration is not a problem. They are used for some surface water sewerage work, but are not satisfactory where the flexibility of the pipeline is important. Porous concrete ogee-jointed pipes can be obtained to BS 1194 for under-drainage purposes.

Pipes of large diameters from 900 to 2600 mm are available with special joints making them suitable for pipe-jacking [90].

Concrete pipes are normally manufactured from ordinary Portland cement, but pipes of sulphate-resisting Portland cement or of high-alumina cement can be obtained where the ground conditions warrant special precautions. CP 2005 states that:

> Sulphate-resisting Portland cement should be used where a moderate concentration of sulphates (from 0·3% to 0·5% of sulphur trioxide in the ground or from 0·1% to 0·3% of sulphur trioxide in the ground water) is in contact with the concrete. Concentrations of sulphur trioxide in the sewage or effluent which occasionally rise to 0·15% would not qualify for consideration as moderate, since there will generally be sufficient dilution to reduce the concentration below 0·1% under conditions of normal flow. It should be noted that Portland type cements are not acid-resistant.

TABLE 4.4
APPROXIMATE WEIGHTS OF CONCRETE PIPES

Nominal bore mm	*Strength class*			
	Std. kg/m	*L kg/m*	*M kg/m*	*H kg/m*
150	50	—	—	—
225	75	—	—	—
300	130	—	130	130
375	180	—	180	140
450	250	—	250	340
575	320	—	420	420
600	430	—	530	530
675	440	—	610	610
750	490	490	610	610
825	720	720	720	720
900	790	790	790	790
975	900	900	900	900
1 050	960	960	960	1 040
1 125	1 090	1 090	1 090	1 090
1 200	1 350	1 350	1 350	1 350
1 350	1 270	1 270	1 500	1 500
1 520	1 900	1 900	1 900	1 900
1 725	1 950	1 950	2 100	2 670
1 800	2 700	2 700	2 700	2 700
1 950	2 300	2 500	2 500	2 600
2 100	2 800	2 800	2 800	3 000
2 400	3 500	3 500	3 500	3 800
2 700	—	4 300	4 300	4 300
3 000	—	5 100	5 100	5 100

By courtesy of the Concrete Pipe Association.

Where the pipes are manufactured from sulphate-resisting Portland cement or from high-alumina cement, the same type of cement should be used in precast manhole sections and in any concrete protection provided to the pipes.

Pipes can be obtained with an inner lining of plasticized PVC sheet to protect the surface of the concrete from hydrogen sulphide gas attack; the lining is applied during pipe manufacture and normally covers 300° to 270° of the surface, the invert remaining unlined.

GRC PIPES

A recent development in concrete pipe manufacture includes the use of glass reinforced cement. This new type of pipe consists of a core of normal concrete with an inner and outer skin of glass reinforced cement. Variations in the thickness of the GRC and of the amount of glass-fibre reinforcement will provide variations in the class of pipe. The pipes are thinner, and therefore lighter, than conventional concrete pipes.

TABLE 4.5
ASBESTOS-CEMENT PIPES—DIMENSIONS AND CRUSHING LOADS (BS 3656)

Nominal diameter mm	*Class L*			*Class M*			*Class H*		
	Ext. diam. mm	*Crushing load*		*Ext. diam. mm*	*Crushing load*		*Ext. diam. mm*	*Crushing load*	
		kgf/m	*kN/m*		*kgf/m*	*kN/m*		*kgf/m*	*kN/m*
100	As class H			As class H			122	3 870	37·95
150	As class H			As class H			176	3 870	37·95
175	As class H			As class H			203	3 870	37·95
200	As class H			As class H			228	3 870	37·95
225	As class H			As class H			254·5	3 870	37·95
250	As class H			As class H			282·5	3 870	37·95
300	As class M			333·5	3 570	35·00	340·5	4 760	46·68
375	As class M			417	3 940	38·63	423	5 360	52·56
450	As class M			495	4 460	43·73	502	5 950	58·35
525	571	3 870	37·95	577	4 840	47·46	585	6 700	65·70
600	651	4 300	42·16	660	5 800	56·87	669	7 900	77·47
675	734	4 900	48·05	743	6 550	64·23	754	8 780	86·10
750	815	5 200	51·00	824	7 000	68·64	836	9 370	91·88
825	897	5 650	55·40	907	7 600	74·53	920	10 270	100·71
900	977	5 950	58·35	993	8 930	87·57	1 003	11 000	107·87
975	1 043	6 400	62·76	1 061	9 670	94·82	1 070	11 900	116·70
1 050	1 123	6 700	65·70	1 141	10 120	99·24	1 154	12 650	124·05

By courtesy of TAC Construction Materials Ltd.

TABLE 4.6
ASBESTOS-CEMENT PIPES—WEIGHTS, kg (BS 3656)

Nom. diam.	*Class L*	*Class M*				*Class H*			
	Pipes (5 m) incl. socket	*Pipes (4 m)*	*Sleeve joints*	*Pipes (4 m) incl. socket*	*Pipes (5 m) incl. socket*	*Pipes (4 m)*	*Sleeve joints*	*Pipes (4 m) incl. socket*	*Pipes (5 m) incl. socket*
mm	*kg*	*kg*	*kg*	*kg*	*kg*	*kg*	*kg*	*kg*	*kg*
100	—	—	—	—	—	37·3	3·4	—	—
150	—	—	—	—	—	63·4	5·2	—	—
175	—	—	—	—	—	78·6	5·5	—	—
200	—	—	—	—	—	99·0	6·3	—	—
225	—	—	—	—	—	109·0	8·7	—	—
250	—	—	—	—	—	133·0	9·5	—	—
300	—	158	11·9	—	—	193	13·1	—	—
375	—	238	19·4	—	—	274	22·0	—	—
450	—	—	—	345	425	—	—	400	492
525	473	—	—	—	538	—	—	—	740
600	611	—	—	—	707	—	—	—	846
675	765	—	—	—	903	—	—	—	1 058
750	936	—	—	—	1 087	—	—	—	1 277
825	1 126	—	—	—	1 312	—	—	—	1 530
900	1 296	—	—	—	1 607	—	—	—	1 815
975	1 495	—	—	—	1 829	—	—	—	2 058
1 050	1 709	—	—	—	2 130	—	—	—	2 410

N.B.—Half-length (2·5 m and 2·0 m) and quarter-length (1·25 m and 1·0 m) are available to order.
By courtesy of TAC Construction Materials Ltd.

ASBESTOS-CEMENT PIPES

Asbestos-cement pipes are manufactured either to BS 486 as pressure pipes suitable for use in rising mains or to BS 3656 for sewerage and drainage gravity pipelines.

The diameters and minimum crushing-test loads for pipes to BS 3656 are given in Table 4.5, while the weights of pipes and joints are given in Table 4.6. These two tables are reproduced by permission from TAC Construction Materials Ltd.

Asbestos-cement pipes are resistant to the substances normally found in domestic sewage and, in general, to wastes considered acceptable for discharge into a sewage treatment plant or a river; they are available with a bitumen coating, if required, for use where fairly high concentrations of sulphate are likely to be encountered or when the pH value of the groundwater is below 6·0. In case of doubt the manufacturer's advice should be sought.

Pipes for sewerage and drainage have plain ends and are flexibly jointed with AC sleeves and rubber rings. Standard lengths are 4 m and 5 m according to diameter, but half-lengths and quarter-lengths are available to order. There is a complementary range of AC junctions, bends, and saddle connections of moulded or fabricated construction.

RPM PIPES

Glass reinforced plastic matrix (RPM) pipes are available in diameters ranging from 750 to 1800 mm, and can be manufactured to order in some other diameters. They are composed of three basic constituents (glass fibre rovings, polyester resin and sand) and are suitable for the conveyance of water and many sewage and industrial effluents.

The pipes are manufactured in four standard classes—6, 10, 12·5 and 16. The class designation expresses the hydraulic working pressure rating of the pipe in bar (N.B.—1 bar = 100 kN/m^2).

Recommendations for the design of buried RPM pipelines, together with notes on handling and installation, are contained in a publication by Stanton and Staveley [95].

PVC PIPES

Plastic pipes have been used for many years in smaller diameters (up to about 50 mm) for water supply pipelines. More recently they have been used quite extensively in the medium diameters (up to 600 mm) for rising mains for both water and sewage. Recent developments in mole-ploughing techniques (see Chapter 12) will no doubt mean a wider use of these pipes as gravity sewers.

Unplasticized PVC (uPVC) pipes are manufactured to BS 3505 and 3506 in diameters up to 600 mm. These pipes are manufactured in four classes B, C, D and E for 'maximum sustained working pressures' of 6·0 bar to 15·0 bar (60 m to 150 m head of water). These pipes are normally supplied in lengths of 3·0, 6·0 and 9·0 m.

BS 4660 relates to 'Unplasticized PVC Underground Drain Pipes and Fittings'. Two nominal diameters are included (110 and160 mm) for standard lengths of 1·0, 3·0 and 6·0 m.

Although uPVC is theoretically a rigid material, from the point of view of structural design of pipelines, uPVC pipes are flexible. Joints are formed either by solvent welding or with collars, using rubber sealing rings. With rubber ring type joints, the collars may be loose, or the ring may fit into an integral socket formed in the pipe itself.

Structural design of flexible pipelines is referred to in Chapter 10, and the additional precautions necessary in bedding are discussed in Chapter 12. The whole subject of handling, laying, jointing and testing PVC pipes is fully dealt with in a PVC Pipelaying Manual [91] issued by a number of leading manufacturers.

When required for use for rising mains, uPVC pipes (to BS 3505) are available to withstand internal hydraulic working pressures as follows:

Class B . . 0·60 MN/m^2 or 60 m head of water
Class C . . 0·90 MN/m^2 or 90 m head of water
Class D . . 1·20 MN/m^2 or 120 m head of water
Class E . . 1·50 MN/m^2 or 150 m head of water

PVC pipes are claimed to be resistant to most acid and alkaline conditions. They are particularly suitable for boggy or made ground and for use in peaty soils. The manufacturers state that the pipes are suitable for depths from 1·00 m to 9·00 m when the bedding and backfilling are correctly carried out.

PITCH-FIBRE PIPES

Pitch-fibre pipes have been manufactured and used in the United Kingdom for about twenty-five years. They are produced to BS 2760 in diameters up to 225 mm and are included on the standard Water Authority estimate form for sewerage schemes. Use of pitch-fibre pipes in this country has, however, generally been limited to small-diameter drainage work.

The pipes are supplied in lengths up to 3 m, and while the majority of manufacturers produce pipes with tapered ends for jointing with push-fit pitch-fibre collars, pipes are obtainable with plain spigot ends for jointing with polypropylene couplings using neoprene 'O' rings. Pitch-fibre pipes can be easily cut on site with a saw, and can then be machined, if necessary, for jointing.

These pipes are claimed to be highly resistant to corrosive chemicals, but the manufacturers should be consulted when their use is contemplated for trade effluent. They are not suitable for

TABLE 4.7
PITCH-FIBRE PIPES—MINIMUM FAILING LOADS AND WEIGHTS

Nominal diameter mm	*Minimum failing load*		*Weight kg/m*
	kgf/m	*N/m*	
50	1 640	16 100	—
75	1 640	16 100	—
100	1 640	16 100	3·4–4·5
125	1 940	19 000	—
150	1 940	19 000	7·3–8·1
200	2 380	23 300	12·1–14·1
225	2 530	24 800	15·2

certain chemicals, such as carbon tetrachloride, hot effluents (over about 60 °C), or for laying at shallow depths. The manufacturers recommend that concrete protection should be given to pipes with less than 450 mm of cover. Attention must be given to correct bedding and backfilling. A technical memorandum on *Pitch-fibre Systems* has been published by the Pitch-fibre Pipe Association [93].

Pitch-fibre drain pipes and perforated pipes are manufactured to BS 2760. Crushing strengths of these pipes (minimum failing loads) are given in Table 4.7, along with weights of pipes, but these latter will vary from one manufacturer to another.

CAST-IRON AND DUCTILE-IRON PIPES

Iron pipes are used where extra strength is required, or for exposed pipes inside buildings. Pressure pipes are used for rising mains; flanged pipes and fittings are generally used inside buildings. Iron pipes are supplied to comply with one of the following British Standards:

BS 437: 'Cast-iron Spigot and Socket Drain Pipes'—for low pressures.
BS 4622: Grey Iron Pipes and Fittings.
BS 4772: Ductile Iron Pipes and Fittings.

Grey iron pipes and fittings are particularly suitable for industrial application where the shock loading is small, weight is unimportant, and surge pressures are minimal. They are available in standard sizes from 80 to 500 mm in two classes. Class 1 pipes are for a maximum hydraulic working pressure of 10 bar (16 bar maximum site test pressure); class 3 pipes for 16 bar maximum working pressure and 25 bar maximum site test pressure.

Ductile iron pipes are available in diameters from 80 to 1200 mm. Recommended maximum working pressures vary, according to diameter and type of joints, from 10 bar to 40 bar. Ductile iron pipes are quick and simple to lay and joint and they are generally suitable for higher working pressures than grey iron pipes.

Mechanical joints incorporating rubber gaskets can be used for all pipes; these joints can be obtained with special stainless steel toothed inserts which are self-anchoring. Flanged pipes and fittings are available in both grey iron and ductile iron.

STEEL PIPES

Steel pipes are rarely used for gravity sewers, as they are more expensive than pipes of the more traditional materials. Steel pipes also need protection against corrosion from sewage. They are used for some inverted siphons and rising mains.

Steel pipes are manufactured to BS 534 and are available in diameters up to 1800 mm (bitumen-lined welded pipes) and up to 300 mm (bitumen-lined seamless pipes). Jointing is normally by Viking-Johnson couplings.

AGRICULTURAL DRAIN PIPES

Clayware field drain pipes to BS 1196 are used extensively for land drainage work. They are available in diameters up to 300 mm and in lengths of 0·30 m. The approximate weights of clayware field drain pipes are set out in Table 4.8.

TABLE 4.8
WEIGHTS OF CLAYWARE AGRICULTURAL DRAIN PIPES

Nominal diameter mm	*Weight per 100 pipes (each 0·30 m) kg*
80	180
100	255
150	455
200	860
225	910
300	1 800

Other pipes used for subsoil drainage include concrete porous pipes to BS 1194, and perforated pipes of clayware, PVC and pitch-fibre.

GRANULAR BEDDING MATERIAL

Granular beddings are by their nature flexible, and where they are used the pipelines should be constructed with mechanical joints. In areas where subsidence may be expected (e.g. mining areas), care should be taken in constructing the joints to ensure that the granular material does not enter the sockets and cause damage to them if subsequent movement takes place. . . The use of granular bedding, particularly with larger diameter extra strength pipes, involves a risk of settlement; correct grading of the bedding material reduces settlement to a minimum. The engineer must consider whether the consequent risk of deviation from line and/or level can be accepted, bearing in mind the purpose and required capacity of the sewer (Ministry Working Party, Second Report).

For small diameter pipes (up to, say, 300-mm diameter) it is normal to use granular material which will pass a 19-mm sieve but be retained on a 4·75-mm sieve. A comparison of the two ISO series of sieves and the original BS series (BS 410) is given in Figure 4.1.

The specification of granular material is also referred to in BS Code of Practice 2005. Gravel and sand, broken stone, and broken brick are satisfactory if the grading is correct, but limestone should not be used in soils containing sulphates or acids. Sharp-edged stones should be avoided with pitch-fibre and PVC pipes, and with specially protected pipes.

For use in fine-graded soils, such as silts, fine sands and mixtures thereof, one part of free-draining coarse sand can be added to two parts of the stone or gravel, and the materials mixed thoroughly, to prevent the intrusion of fine-grained materials from the neighbouring soil.

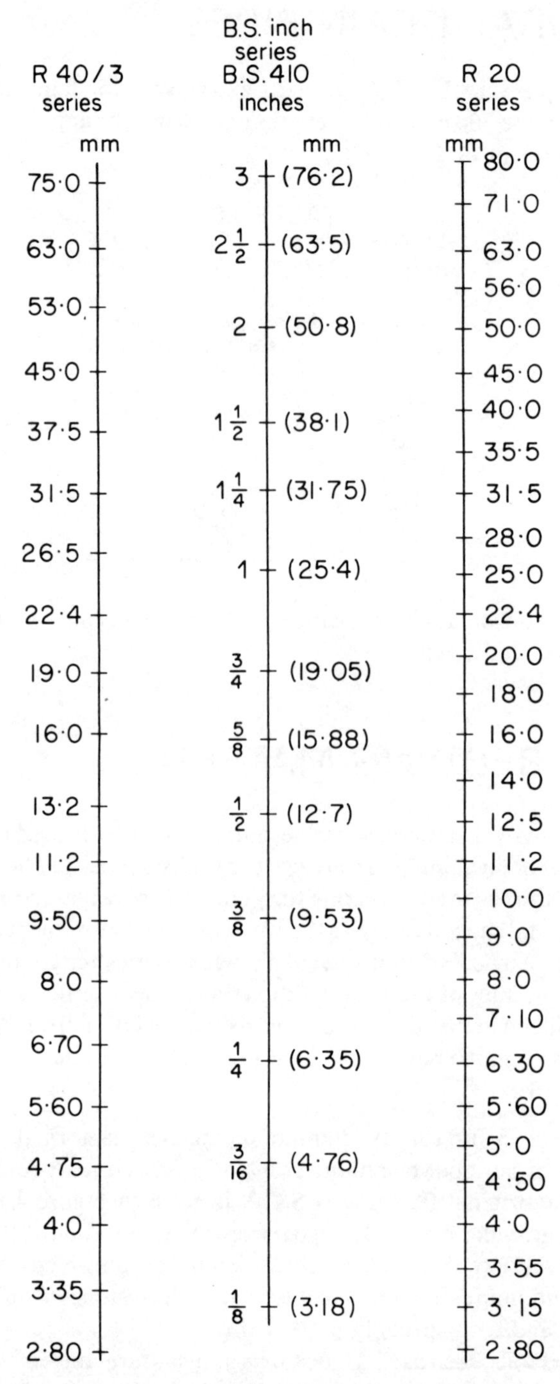

Fig. 4.1. Comparison of some sieve sizes in the 1962 *edition of BS* 410 *with the two ISO Series.*

MANHOLES AND INSPECTION CHAMBERS

As an alternative to *in situ* manholes of brick or concrete, some engineers use specially manufactured precast concrete units. These can be obtained to BS 556 and comprise base units, chamber rings, taper units, shaft rings, and cover slabs. Tapers and shafts are normally used when the depth to invert is more than about 2·5 or 3 m (see Figure 11.2). The main chamber rings are available in diameters ranging from 900 to 2000 mm and while the shaft rings were normally 700-mm diameter, many authorities are now specifying shafts of 900-mm diameter. Chamber rings, tapers and shaft rings are supplied fitted complete with stepirons to BS 1247.

Similar precast units are available for the construction of house inspection chambers. These are supplied in diameters of 600, 750 or 900 mm to BS 556, complete with a concrete cover and frame, the latter forming a reducing slab over the circular chamber rings. Plastic inspection chamber units are available for use with either plastic or vitrified clay pipes. The manufacturers of plastic chambers state that they are strong enough to support BS 497 grade 'C' loading without the use of a concrete surround.

Manhole covers and frames are specified in BS 497. For service in roads they should be grade 'A' triangular or double triangular type, but where only light loading is expected they can be to grade 'B' (circular). A variety of patented 'non-rocking' covers is also available in either the triangular or 'cloverleaf' type for class 'A' loading. Class 'C' rectangular covers and frames to BS 497 should not be used in places accessible to vehicles, and these are generally only used over inspection chambers on house drainage. The 1976 edition of BS 497 extended the standard so that the format is now that of a performance specification with minimal dimensional control; the standard also now includes ductile iron (spheroidal graphite iron) alongside conventional grey iron. Ductile iron covers and frames are very much lighter in weight.

In addition to 'standard' covers, a large number of variations is available. Covers can be recessed for filling with concrete, etc.; they can be complete with ventilation slots; they can be watertight or pressure tight; and they can be lockable. Ventilating covers are now rarely used, and where a manhole or inspection chamber is sited within a building the cover should be watertight, the frame fitted with an airtight seal, and the cover secured to the frame by removable bolts.

VALVES AND PENSTOCKS

Valves can be obtained to BS 1218, 'Sluice Valves for Waterworks Purposes', or to BS 3464, 'Cast-iron Gate Valves for General Purposes'. While valves to the former specification were often specified in the past, it is more logical to specify valves to BS 3464 for sewerage work. They should be fitted with a cleaning door or an access plug, and should have 'external screw' spindles. They can be operated by tee-key or handwheel, or through a headstock. Where required, a locking device can be incorporated. When ordering valves, the working and maximum pressure should be quoted. Penstocks will vary according to the duty and whether the pressure is 'on-seating' (tending to force the door on to the frame) or 'off-seating' (tending to force the door away from the frame). While the former is preferable, this may not always be possible. The choice of material for the sealing faces will depend on the frequency of operation and whether the liquid is of a corrosive nature. Screw-type penstocks for sewerage work should preferably have gunmetal faces on both frame and gate, and should be fitted with screwed forged bronze spindles working through gunmetal nuts. Various forms of disc and weir penstocks are available for special duties.

Reflux valves should be designed to avoid reverse flow and the consequent slam on the seating when closing. They should be specified for use with sewage. To avoid deposition of solids on the seating, reflux valves on sewage rising mains should always be fixed on a horizontal section of pipeline. Doors and door seatings should be of gunmetal, while the hinge pins should be of stainless steel.

Fig. 4.2. An air release valve designed for use with sewage (sectioned example by courtesy of Glenfield & Kennedy Ltd).

Air valves for sewage rising mains should be specially designed for use with sewage and should preferably have both large and small orifices. This ensures that large volumes of air are quickly released while the main is being filled; small accumulations of air or gas can be released while the main is under pressure; and sufficient air is admitted to the main when the pressure falls below atmospheric pressure (see Fig. 4.2).

CONCRETE

Cement is delivered to the site in 50-kg bags or in bulk by the tonne. Portland cement manufactured to BS 12 is used for most concreting work on sewerage contracts. Where sulphates are present in solution in the ground water or in clay soils, or where they are likely in the sewage, the concrete should be made from sulphate-resisting Portland cement to BS 4027 or from supersulphated cement to BS 4248. Good dense concrete made with sulphate-resisting Portland cement can be proof against 0·5 % sulphates (as sulphur trioxide, SO_3) in ground water or 2·0 % by weight in clay. The conditions for using these cements are set out fully in CP 2005.

Aggregates should comply with BS 882 and 1201, graded to produce a concrete of the specified quality. Water should comply with BS 3148 ('Tests for Water for Making Concrete'). The water/cement ratio is important, particularly for a water-retaining structure. Other things being equal, a low ratio (and therefore low workability) gives a stronger concrete, less liable to shrinkage cracks.

While nominal volume mixes have been common in the past, it is now usual to specify concrete mixes either by weight or according to the strength required. The use of designed mixes is encouraged, the mix depending on requirements of strength, workability, durability and impermeability.

Both nominal and designed concrete mixes are considered in BSCP 114, 'The Structural Use of Reinforced Concrete in Buildings' and in BSCP 110, 'The Structural Use of Concrete'. A booklet published by the Institution of Civil Engineers entitled *A Guide to Specifying Concrete* [38] refers to materials, mix design, transport and placing, and formwork. Table 4.9, reproduced from that booklet with the permission of the Council of the Institution, sets out the minimum times before removal of soffit forms and props, in the absence of cube-strength results. It is recommended that the formwork to vertical surfaces may be removed after twelve hours (when the temperature is over 15 °C).

Useful Ministry of Public Building and Works leaflets on the subject of concreting are:

No. 7. 'Concreting in Cold Weather'.
No. 26. 'Making Concrete'.
No. 39. 'Cements Other than Ordinary Portland Cement'.
No. 60. 'Ready Mixed Concrete'.

All concrete should be properly cured to obtain the full advantage of quality control. The 'setting' of concrete is part of the process of the hydration of the cement, and any excessive loss of moisture through evaporation at that stage will entail a loss in strength of the concrete. Concrete

TABLE 4.9
MINIMUM TIMES FOR THE REMOVAL OF SOFFIT FORMS (IN DAYS)

Location	*Ordinary Portland cement concrete*		*Rapid-hardening Portland cement concrete*	
	Cold weather, 2°–5°C	*Normal weather, about 15°C*	*Cold weather, 2°–5°C*	*Normal weather, about 15°C*
Slab soffit forms (props left under)	10	3	8	2
Beam soffit forms (props left under)	14	7	9	4
Props to slabs	21	7	11	4
Props to beams	28	16	21	8

With acknowledgement to The Council of The Institution of Civil Engineers.

surfaces should therefore be protected from the sun and drying winds for at least the first 14 days, and this should preferably be extended up to about 28 days, depending on the weather conditions. Curing can be achieved by regular spraying, by ponding (flat surfaces), by the use of sheets of various materials, or by covering with damp sand. Proprietary curing membranes are available, which rely on their sealing properties to retain the initial moisture.

When a granolithic finish is required, this may either be monolithic with the main slab (placed within 2 or 3 hours) or laid separately. A monolithic finish can be about 20 mm in thickness, but if the topping concrete is laid after the base has hardened, the topping should be about 40 mm thick. Aggregates should comply with BS 1201, and should be of igneous rock; coarse aggregate should have a maximum size of 9·5 mm, mainly retained on a 4·75-mm sieve; fine aggregate should mainly pass a 4·75-mm sieve. In general, a 3:1 mix by weight is used with a water/cement ratio not more than 0·42. The topping must be properly compacted and adequately trowelled. Curing is a vital factor in obtaining a good quality concrete free from shrinkage cracks.

TABLE 4.10

STEEL REINFORCEMENT—AREA FOR A GIVEN NUMBER OF BARS IN mm²

Diam. mm	*Number of bars*									
	1	*2*	*3*	*4*	*5*	*6*	*7*	*8*	*9*	*10*
6	28	56	84	113	141	169	198	226	254	283
8	50	100	150	201	251	301	352	402	452	503
10	78	157	235	314	392	471	549	628	706	785
12	113	226	339	452	565	678	791	904	1 017	1 131
16	201	402	603	804	1 005	1 206	1 407	1 608	1 809	2 011
20	314	628	942	1 256	1 571	1 885	2 199	2 513	2 827	3 142
25	490	981	1 472	1 963	2 454	2 945	3 436	3 927	4 418	4 909
32	804	1 608	2 412	3 216	4 021	4 825	5 629	6 433	7 237	8 042
40	1 256	2 513	3 769	5 026	6 283	7 539	8 796	10 052	11 309	12 566
50	1 963	3 927	5 890	7 854	9 817	11 781	13 744	15 708	17 671	19 635

By courtesy of Reinforcement Steel Services, Sheffield.

TABLE 4.11

STEEL REINFORCEMENT—AREA PER METRE WIDTH FOR VARIOUS BAR SPACINGS (IN mm²)

Diam. mm	*Spacing of bars in mm*								
	50	*75*	*100*	*125*	*150*	*175*	*200*	*250*	*300*
6	566	377	283	226	188	161	141	113	94
8	1 006	670	503	402	335	287	251	201	167
10	1 570	1 046	785	628	523	448	392	314	261
12	2 262	1 508	1 131	904	754	646	565	452	377
16	4 022	2 681	2 011	1 608	1 340	1 149	1 005	804	670
20	6 284	4 189	3 142	2 513	2 094	1 795	1 571	1 256	1 047
25	9 818	6 544	4 909	3 926	3 272	2 805	2 454	1 963	1 636
32	16 084	10 722	8 042	6 433	5 361	4 595	4 021	3 216	2 680
40	—	—	12 566	10 052	8 377	7 180	6 283	5 026	4 188
50	—	—	19 635	15 708	13 090	11 220	9 817	7 854	6 545

By courtesy of Reinforcement Steel Services, Sheffield.

TABLE 4.12
STEEL REINFORCEMENT—MASS FOR A GIVEN NUMBER OF BARS IN kg/m RUN

Diam. mm	*Number of bars*									
	1	*2*	*3*	*4*	*5*	*6*	*7*	*8*	*9*	*10*
6	0·222	0·444	0·666	0·888	1·110	1·332	1·554	1·776	1·998	2·220
8	0·395	0·790	1·185	1·580	1·975	2·370	2·765	3·160	3·555	3·950
10	0·616	1·232	1·848	2·464	3·080	3·696	4·312	4·928	5·544	6·160
12	0·888	1·776	2·664	3·552	4·440	5·328	6·216	7·104	7·992	8·880
16	1·579	3·158	4·737	6·316	7·895	9·474	11·053	12·632	14·211	15·790
20	2·466	4·932	7·938	9·864	12·330	14·796	17·262	19·728	22·194	24·660
25	3·854	7·708	11·562	15·416	19·270	23·124	26·978	30·832	34·686	38·540
32	6·313	12·626	18·939	25·252	31·565	37·878	44·191	50·504	56·817	63·130
40	9·864	19·728	29·592	39·456	49·320	59·184	69·048	78·912	88·776	98·640
50	15·413	30·826	46·239	61·652	77·065	92·478	107·891	123·304	138·717	154·130

By courtesy of Reinforcement Steel Services, Sheffield.

REINFORCEMENT

BS 4449 relates to hot rolled steel bars and hard-drawn steel wire; BS 4461 relates to cold worked steel bars; and BS 4483 to steel fabric reinforcement. All steel should be free from mill scale and loose rust, grease, etc., and the concrete cover to all steel (*including* stirrups) should not be less than 38 mm. During recent years there has been an increasing use of high-tensile steel in preference to mild-steel bars.

Economy in the office and on site can be obtained by limiting the number of bent shapes to the preferred shapes set out in BS 4466, 'Bending Dimensions and Scheduling of Bars for the Reinforcement of Concrete', and by standardizing bar-bending schedules.

Sectional areas of bars, areas of steel per metre width and weights of steel bars are given in Tables 4.10, 4.11 and 4.12. BS preferred types of steel fabric are given in Table 4.13.

BRICKWORK AND BLOCKWORK

CP 2005 recommends that bricks for lining sewers and inverts should preferably comply with BS 3921, 'Bricks and Blocks of Fired Brickearth, Clay or Shale'. The standard brick dimensions of 215 × 102·5 × 65 mm, together with mortar joints of 10-mm thickness, give a working brickwork format size of 225 × 112·5 × 75 mm (see BS 3921, Part II).

Precast concrete blocks to BS 2028 and 1364, 'Specification for Precast Concrete Blocks', are of three types. Types A and B are suitable for use below damp-proof course level. A number of manufacturers produce blocks 200 mm wide by 400, 500 and 600 mm long, with thicknesses of 100 mm and other dimensions to meet performance requirements. The standard size for precast concrete facing blocks is 200 × 100 × 75 mm nominal; other sizes are produced. BS 2028 specifies densities and compressive strengths as set out in Table 4.14. Lightweight load-bearing insulating building blocks to type 'B' of the BS are used extensively to provide the fire resistance and insulation required by the Building Regulations.

TABLE 4.13
BS PREFERRED TYPES OF STEEL FABRIC

Ref.	Nominal pitch mm		Size of wires mm		Cross-sectional area per metre width, mm^2	
	Main	Cross	Main	Cross	Main	Cross
A393	200	200	10	10	393	393
A252	200	200	8	8	252	252
A193	200	200	7	7	193	193
A142	200	200	6	6	142	142
A98	200	200	5	5	98	98
B1 131	100	200	12	8	1 131	252
B785	100	200	10	8	785	252
B503	100	200	8	8	503	252
B385	100	200	7	7	385	193
B283	100	200	6	7	283	193
B196	100	200	5	7	196	193
C785	100	400	10	6	785	70·8
C636	100	400	9	6	635·9	70·8
C503	100	400	8	5	503	49·0
C385	100	400	7	5	385	49·0
C283	100	400	6	5	283	49·0
D49	100	100	2·5	2·5	49·1	49·1
D31	100	100	2·0	2·0	31·4	31·4

With acknowledgements to the British Standards Institution.

TABLE 4.14
PRECAST CONCRETE BLOCKS TO BS 2028 AND 1364

Block type	Density of block kg/m^3	Minimum average compressive strength N/mm^2	Strength—lowest individual block N/mm^2
A	Not less than 1 500	3·5 7·0 10·5 14·0 21·0 28·0 35·0	2·8 5·6 8·4 11·2 16·8 22·4 28·0
B	Less than 1 500, but more than 625	2·8 7·0	2·25 5·6
	Less than 625	2·8	2·25
C	Less than 1 500, but more than 625	Transverse breaking load is specified (varies with size of block)	
	Less than 625		

With acknowledgements to the Cement and Concrete Association.

A set of safe load tables for concrete blockwork to the design requirements of CP 111 has been published by the Cement and Concrete Association [96].

TIMBER

When a pitched-roof construction is proposed for a pumping station building, timber is generally employed for the roof members. Timber is, of course, also used in piling and in concrete formwork.

The structural use of timber is controlled by CP 112, Part 2, while BS 4978 specifies the various timber grades for structural usage.

Hardwoods are available in thicknesses from 19 to 100 mm, and in widths from 50 to 300 mm (see BS 5450). Other dimensions are available on request. Softwoods are available in various basic lengths from 1·80 to 6·30 m and in the basic sizes given in Table 4.15. It should be noted that the Timber Trade Federation has stated that the 36- and 40-mm thicknesses may not be readily available. A further thickness of 47 mm may be available in some localities.

Stress-graded timber (to BS 4978) is referred to in CP 112 and also in Schedule 6 of the Building Regulations.

TABLE 4.15
BASIC SIZES OF SAWN SOFTWOOD (CROSS-SECTIONAL SIZES)—BS 4471

Thickness in mm	*Width in mm*								
	75	*100*	*125*	*150*	*175*	*200*	*225*	*250*	*300*
16	×	×	×	×	—	—	—	—	—
19	×	×	×	×	—	—	—	—	—
22	×	×	×	×	—	—	—	—	—
25	×	×	×	×	×	×	×	×	×
32	×	×	×	×	×	×	×	×	×
36	×	×	×	×	—	—	—	—	—
38	×	×	×	×	×	×	×	—	—
40	×	×	×	×	×	×	×	—	—
44	×	×	×	×	×	×	×	×	×
50	×	×	×	×	×	×	×	×	×
63	—	×	×	×	×	×	×	—	—
75	—	×	×	×	×	×	×	×	×
100	—	×	—	×	—	×	—	×	×
150	—	—	—	×	—	×	—	—	×
200	—	—	—	—	—	×	—	—	—
250	—	—	—	—	—	—	—	×	—
300	—	—	—	—	—	—	—	—	×

By courtesy of the Timber Trade Federation.

STRUCTURAL STEEL

BS 4, Part 1 (with amendments) gives the dimensions and other properties of Hot Rolled Sections, together with details of rolling tolerances. Safe Load Tables have been incorporated into a

Handbook on Structural Steel [98] published by the British Constructional Steelwork Asoociation in 1971.

All surfaces of structural steel members should be given a coat of red lead paint before assembly at the manufacturer's works. Any steel delivered to the site unpainted should be cleaned and painted with a coat of red lead paint as early as possible. Site bolts and rivets should be painted after erection, and any damage to the priming coat should then be touched up. Surfaces to be welded on site or surfaces to be joined with high strength friction grip bolts should, however, be left unpainted.

STEEL PLATES FOR FLOORING

Two types of flooring plates are in common use for pumping stations and similar buildings—open-type and solid plating with raised non-slip patterns.

TABLE 4.16
OPEN-TYPE STEEL FLOORING—SAFE LOADING kg/m² (STEELWAY TYPE 1 FLOORING)

Depth of flooring mm	Straight bar and (pressed bar) mm	Weight in kg/m²	Span in mm											
			600	700	800	900	1 000	1 200	1 400	1 600	1 800	2 000	2 200	2 400
20	20 × 3 (20 × 3)	28·8	2 090	1 519	1 020	714	520	337	—	—	—	—	—	—
25	25 × 3 (20 × 3)	32·4	3 314	2 437	1 866	1 377	999	581	367	—	—	—	—	—
25	25 × 5 (20 × 3)	41·0	4 905	3 600	2 753	2 050	1 499	867	551	367	—	—	—	—
35	35 × 3 (25 × 3)	43·1	6 281	4 619	3 538	2 794	2 264	1 530	959	642	449	—	—	—
35	35 × 5 (25 × 3)	55·2	9 402	6 903	5 292	4 181	3 385	2 315	1 458	979	683	500	—	—
40	40 × 5 (25 × 3)	60·8	12 033	8 831	6 761	5 343	4 334	3 008	2 131	1 428	999	734	500	—
45	45 × 5 (25 × 3)	66·3	14 990	11 013	8 433	6 669	5 405	3 753	2 753	2 009	1 407	1 030	704	500
50	50 × 5 (25 × 3)	71·8	18 355	13 460	10 299	8 148	6 598	4 579	3 365	2 580	1 917	1 397	958	673

Deflection limited to span/200 or 10 mm whichever is lesser.
Loads shown to the right of the heavy line are below the minimum requirements of BS 4592: 1970.
1 kN/m² = 101·9718 kg f/m² = 20·8854 lb f/ft².
By courtesy of Steelway (Glynwed Integrated Services Ltd).

TABLE 4.17
DURBAR FLOOR PLATES—DIMENSIONS AND WEIGHTS

Thickness on plain mm	Standard length mm	Standard width mm	Weight kg/m²
4·5	4 000	1 750	37·97
6	4 000	1 750	49·74
8	4 000	1 750	65·44
10	6 000	1 750	81·14
12·5	6 000	1 750	100·77

By courtesy of the British Steel Corporation.

TABLE 4.18
DURBAR PATTERN STEEL PLATES: SAFE LOADINGS—SIMPLY SUPPORTED ON TWO SIDES, kg/m²

Thickness on plain mm	*Span in metres*							
	0·6	*0·8*	*1·0*	*1·2*	*1·4*	*1·6*	*1·8*	*2·0*
4·5	1 228	697	447	310	228	176	137	112
6·0	2 204	1 240	794	552	404	312	244	198
8·0	3 920	2 210	1 408	982	717	554	434	356
10·0	6 116	3 441	2 198	1 532	1 121	866	677	555
12·5	9 571	5 385	3 439	2 397	1 754	1 355	1 060	869

By courtesy of the British Steel Corporation.

TABLE 4.19
DURBAR PATTERN STEEL PLATES: SAFE LOADINGS—SIMPLY SUPPORTED ON FOUR SIDES, kg/m²

Thickness on plain mm	*Span in m*								*Breadth m*
	0·6	*0·8*	*1·0*	*1·2*	*1·4*	*1·6*	*1·8*	*2·0*	
4·5	2 483	1 634	1 403	1 319	1 284	1 269	1 256	1 251	0·6
		1 396	985	937	772	742	726	716	0·8
			894	662	564	515	490	475	1·0
6·0	4 408	2 899	2 490	2 342	2 279	2 246	2 230	2 221	0·6
		2 479	1 747	1 485	1 371	1 317	1 288	1 272	0·8
			1 587	1 176	1 000	914	869	843	1·0
				1 102	849	725	660	622	1·2
8·0	7 836	5 154	4 427	4 163	4 052	3 993	3 965	3 950	0·6
		4 408	3 106	2 638	2 438	2 341	2 291	2 260	0·8
			2 821	2 089	1 778	1 625	1 544	1 498	1·0
				1 958	1 510	1 288	1 173	1 107	1·2
					1 438	1 142	983	893	1·4
10·0	12 250	8 054	6 916	6 504	6 330	6 238	6 194	6 170	0·6
		6 887	4 853	4 123	3 809	3 658	3 580	3 532	0·8
			4 408	3 265	2 779	2 538	2 413	2 342	1·0
				3 060	2 359	2 013	1 832	1 729	1·2
					2 248	1 784	1 536	1 394	1·4
12·5	19 140	12 590	10 810	10 160	9 893	9 750	9 681	9 642	0·6
		10 760	7 584	6 443	5 953	5 716	5 594	5 520	0·8
			6 889	5 102	4 343	3 967	3 770	3 659	1·0
				4 783	3 687	3 147	2 864	2 702	1·2
					3 513	2 789	2 400	2 179	1·4

By courtesy of the British Steel Corporation.

Open type flooring is manufactured in various types and in depths to suit the loading. Table 4.16 gives the weights and safe loadings for Steelway open-type flooring.

The 'Admiralty' diamond pattern steel flooring which was used for many years has been superseded by the 'Durbar' pattern plates which are self-draining and easier to clean. Table 4.17 gives lengths and weights of Durbar pattern plates, while Tables 4.18 and 4.19 give safe uniformly distributed loads.

GLAZING

The subject of glazing is covered very fully in a *Glazing Manual* published by the Glass and Glazing Federation [87]. This deals with the types of glass available, glazing materials and techniques, and general information on design.

TABLE 4.20
SHEET GLASS

Nominal thickness mm	*Thickness range mm*	*Approx. mass kg/m²*	*Normal maximum size mm*
3	2·8–3·2	7·5	2 130 × 1 230
4	3·7–4·3	10·0	2 760 × 1 220
5	4·7–5·3	12·5	2 130 × 2 400
6	5·7–6·3	15·0	2 130 × 2 400

Three qualities are available:

OQ— ordinary glazing quality—for general purposes in factories, housing estates, etc.;

SQ— selected glazing quality—for glazing in buildings requiring a better quality of glass;

SSQ— special selected quality—for high grade work such as pictures, cabinet work, etc.

By courtesy of the Glass and Glazing Federation.

TABLE 4.21
FLOAT OR POLISHED PLATE GLASS

Nominal thickness mm	*Thickness range mm*	*Approx. mass kg/m²*	*Normal maximum size mm*
3	2·8–3·2	7·5	2 140 × 1 220
4	3·8–4·2	10·0	2 760 × 1 220
5	4·8–5·2	12·5	3 180 × 2 100
6	5·8–6·2	15·0	4 600 × 3 180
10	9·7–10·3	25·0	6 000 × 3 300
12	11·7–12·3	30·0	6 000 × 3 300
15	14·5–15·5	37·5	3 050 × 3 000
19	18·0–20·0	47·5	3 000 × 2 900
25	24·6–26·0	63·5	3 000 × 2 900

By courtesy of the Glass and Glazing Federation.

Sheet glass is available in the thicknesses and sizes set out in Table 4.20, while float glass sizes are given in Table 4.21. For pumping stations and similar buildings it would be normal to use ordinary glazing quality (OQ) sheet glass or float glass. The thickness of glass should be calculated according to the wind loading to be expected (suitably corrected for height above ground) and the type of glass to be used.

Various types of translucent and other glasses are available, and float glass can be obtained either clear or tinted. Polished wired glass is availablc in 6-mm thickness.

5 Contract and Site Organization

CONTRACTS and site organization, particularly the supervision of civil engineering works, are dealt with very fully in a book by Twort [76]. This chapter stresses some of the points made in that book, draws on the author's own experience as a resident engineer, and makes reference to the various codes of practice and other relevant publications.

The 'engineer' to a contract may be a direct employee of the 'employer' (e.g. a local authority engineer), or he may be an independent consultant. Whichever is the case, the engineer has two quite separate tasks. First, he advises the employer on the design of the works, and he prepares drawings, specifications, bills of quantities, etc. Secondly, once a contract has been entered into for the construction of the works, he must also act as an independent arbitrator between the employer and the contractor on all contractual matters.

This chapter is mostly concerned with the method of construction where a contractor has been appointed, based on a competitive tender, and it will not therefore be relevant in some respects to lump-sum contracts or to works carried out by direct labour.

The design and the construction of sewers are very interdependent; a knowledge of each of these aspects of sewerage is a prerequisite to obtaining a satisfactory completed scheme. The efficiency and the economy of the scheme will be affected as much by the specification and the standards of site control as by the original design.

It will be apparent that the design engineer needs site experience to guide him when preparing a specification, and equally that the engineer supervising construction (usually referred to as the 'resident engineer'), must have had design experience if he is to appreciate the reasons for some of the clauses in the specification.

With the increasing shortage of skilled labour, the contractor is tending to rely more on mechanization during construction. Both the engineer responsible for the design and specification and the resident engineer must make allowance for this increase in use of mechanical plant. The Ministry of Housing and Local Government Second Report, referring to the use of modern construction methods, points out that the more the engineer can cater for these 'the more easily and cheaply he will attain the required quality of finished product, the standard of which must not be allowed to and need not deteriorate as a consequence of mechanized construction'.

CODES OF PRACTICE

During the last few years many new BS codes of practice have been published. These codes are not specifications, and are not therefore normally referred to in the contract documents. A code of practice is, however, intended to indicate what is considered by the drafting committee to be good practice in design and construction, and to summarize information on the subject for the benefit of engineers engaged on either design or construction.

BSCP 2005, 'Sewerage', and BSCP 301, 'Building Drainage', are the principal codes relating to sewerage and drainage works; other codes (listed in Appendix B) have some relevance. The sewerage code of practice has sections dealing with materials and components, fundamental considerations, design and construction, ancillary structures, pumping stations and rising mains, and tidal outfalls.

Probably almost as important to the engineer working on a sewerage contract is BSCP 110 which deals with the 'Structural Use of Concrete'. This Code fully covers materials, design and construction of reinforced concrete works, including formwork, while BS 5337 relates to the 'Structural Use of Concrete for Retaining Aqueous Liquids'.

During the early stages of a contract, CP 2001, 'Site Investigations', will prove a valuable guide to the soils and rocks encountered—their definition, sampling and field-testing. CP 2003, 'Earthworks', develops this subject and deals with earthwork construction, the design and construction of cuttings, embankments, trenches and pits, cut and fill, compaction, protection of slopes and excavation in rock.

Other useful BS Codes which should be available in the site office are CP 2010: Part 1, 'Installation of Pipelines in Land', CP 303, 'Surface Water and Subsoil Drainage', and CP 2004 'Foundations'. CP 2004 contains valuable guidance on bearing capacities and pressures, spread and piled foundations, cofferdams, caissons, demolitions and dewatering; the Code also has a section on tidework, underwater concreting and diving.

CONDITIONS OF CONTRACT

Various standard forms of conditions of contract are obtainable to cover civil engineering, mechanical and electrical engineering (with or without erection), architectural work, etc. These are published by the relevant professional bodies or by government departments. Very few consultants or local authorities now have their own 'conditions', and for most civil engineering contracts (including sewerage) they use the standard *General Conditions of Contract and Forms of Tender, Agreement and Bond for Use in Connection with Works of Civil Engineering Construction*, prepared in 1950 by the Institution of Civil Engineers jointly with the Association of Consulting Engineers and the Federation of Civil Engineering Contractors (as subsequently amended).

Conditions of contract set out the general obligations of both parties to the contract and contain clauses covering a wide range of subjects which could otherwise be liable to dispute during the contract. These range from insurance of the works, setting out, accidents on site, programme, etc., to measurements and payments, nominated sub-contractors and the action to be taken in case of outbreak of war.

This document relates to general contractual rights and duties, and does not refer to specific standards of materials or workmanship (which are covered by the specification), nor to quantities (which are set out in the bills of quantities). The 5th Edition (1973) of the ICE *Conditions of Contract* is at present the current issue.

It is usual for a set of contract documents to include a number of 'special conditions', in addition to those set out in the general ICE *Conditions*. These will vary from contract to contract and will cover such subjects as specific definitions of 'employer', 'engineer' and 'site', and any particular provisions, such as recruitment of labour and co-operation with other contractors. For overseas schemes, the complete conditions of contract are normally prepared on a 'one-off' basis based on local standards.

Clause 52 (3) of the ICE *Conditions* refers to the *Schedules of Dayworks carried out incidental to Contract Work* which are published by the Federation of Civil Engineering Contractors. These schedules contain hourly or daily rates for plant and equipment and quote the percentage additions applicable on labour rates and the cost of materials. They are, of course, revised from time to time, and it is important to have the most up-to-date edition on a contract.

Clause 57 of the *Conditions* refers to the *Standard Method of Measurement of Civil Engineering Quantities* issued by the Institution of Civil Engineers. Where measurement in the bills of quantities does not follow the recommendations of that document, the difference in the method used must be clearly stated.

SPECIFICATION

While the conditions of contract can be standardized to a very great extent, the specification should not be stereotyped. This has already been referred to in Chapter 4. Some clauses will, of course, be repeated from previous contracts, but the document as a whole must be prepared with the specific scheme in mind and must be kept up-to-date in the light of experience.

The structural design of pipelines includes a safety margin (see Chapter 10). While the Ministry Working Party has proposed a value of 0·80 for this, many engineers are of the opinion that this figure must be subject to review for each contract, in the light of site conditions and based on the standards of both contractor and supervision. Clarke [67] has suggested:

> providing for alternative beddings in the specification, and calling for tenders on the alternatives, whilst ordering a suitable class of pipe for the lower bedding class and a normal factor of safety, before the contract is advertized. The decision as to which bedding to adopt may then be left until the tenders are received when the weighting of doubtful low bids by the more expensive bedding may produce a more equitable comparison of the tenders and a safer pipeline.

The Ministry Working Party Second Report refers to the practice in USA for the engineer to decide what different pipe materials are acceptable, to specify the bedding and class of pipe required with each material, and to allow the contractor to offer whichever material he wishes. They recommend that the engineer should:

> specify precisely what is required with each acceptable combination but to allow the contractor to tender for that specified combination which his skill, experience, plant and labour force will allow him to construct most cheaply and efficiently. This system would create true competition both in the construction and pipe supply fields.

The specification should contain clauses to cover all types of materials and workmanship envisaged. Additional clauses will be included to describe the required standards for general items, such as the engineer's office, toilet facilities on site, access to site, fencing, limitations on advertizing, and clearing the site on completion.

Drawings and specifications must always be considered as supplementary to each other. A good set of documents will state clearly by word or drawing the nature and extent of the work to be performed, together with details of acceptable materials and methods of construction. The bills of quantities must include sufficient items, so that the tenderer can determine his prices sufficiently accurately; this should then reduce subsequent claims and disputes to the minimum.

INSURANCE

The ICE *Conditions of Contract* contain certain clauses which require the contractor to take out insurance cover. This subject is generally covered in Clauses 21 to 25 and relates to insurance of the works and temporary works (with certain 'excepted risks') together with separate insurances to cover public liability and employers' liability risks.

It is essential that the insurance cover should be arranged for and kept in force during the whole period of the contract, including the period of maintenance. The policies for works and public liability must be taken out in the joint names of the contractor and the employer, and it is accepted practice for the employer to examine the appropriate policies and to ensure that the premium payments are kept up-to-date.

It should, however, be borne in mind that once a certificate of completion has been issued for all or a part of the works, and the period of maintenance has commenced, *the contractor's liability* is reduced to loss or damage which arise from a cause which occurred prior to the period of maintenance, or due to work actually carried out by him during that period. Damage occurring during the period of maintenance is not the responsibility of the contractor unless it is due to his negligence.

On a large contract it will not be necessary to have insurance cover for the full contract sum in the early stages. The rate of build-up will depend on the type of contract and should allow for any rapid increase in value due to delivery of machinery, etc. The actual premiums payable by the contractor would, however, normally be quoted as a percentage of the contract price.

As with motor and other insurance, an 'excess' figure is often applied to this type of insurance. The actual figure will depend on the type of claim and on the individual policy. As this is a joint policy, the amount of excess must, of course, be agreed by the employer, even though it is carried by the contractor.

While the 'excepted risks' are defined in Clause 20 (2) of the ICE *Conditions of Contract*, on certain contracts it may be relevant to adjust these to take account of local conditions.

The minimum amount of third party (public liability) insurance required under Clause 23 (2) of the *Conditions of Contract* is to be stated by the employer at the appendix to the form of tender. This is a *minimum* figure for any one accident (the total amount during the period of cover being unlimited), and there is no reason why the contractor should not increase this figure if the contract conditions appear to be particularly hazardous.

The requirements of Clause 24 are normally met by the contractor's annual employer's liability insurance.

PROGRAMME

While the contract documents will often include an overall programme for the guidance of tenderers, the successful contractor will usually be required to draw up his own proposed programme on the basis of the availability of his labour, plant and materials. With this programme available 'as soon as practicable after the acceptance of the tender' (Clause 14 of the *Conditions*), an estimate can be made of the employer's financial obligations during the contract, and provision can be made for the timing of subsequent contracts (e.g. house connections). This early programme must be elaborated or modified as the contract progresses and must be regularly compared with progress actually made.

A sewerage contract can usually be flexible to some extent, and the programme of work should be drawn up, bearing in mind the interests of all parties concerned, particularly where a pipeline will pass through private property.

> As much notice as possible of intention to enter upon the land for construction or any subsequent maintenance or relaying work should be given to individual owners and occupiers and to any other authority affected. The implications of the work and the programme should be discussed with the occupiers, and arrangements made for suitable access to the pipeline route for personnel and equipment. . . . The occupier should be advised whether the work will be completed in one operation or whether a return will be necessary for other works after the main pipelaying (CP 2010).

All civil engineering contracts are being increasingly mechanized. The high cost of mechanical plant makes it essential for the contractor to be able to work continuously, despite bad ground or weather conditions. It is therefore the duty of the engineer to design sewers for which modern constructional facilities can be fully employed, and to cater for the limitations imposed by mechanization.

The programme itself can be drawn up as a combined 'programme and progress' bar chart for smaller schemes. For larger and more important contracts, the critical path method (CPM) may be more suitable. Accepted methods of maintaining progress records by bar charts and 'financial' charts are described by Twort [76] and are also referred to in Ministry of Works leaflets.

The critical path method of analysis is explained in a book by Lockyer, *An Introduction to Critical Path Analysis* [74]. Very briefly, the usual approach to the critical path method is in three stages:

1. To list all activities, and then to draw the network showing them in their proper sequence.
2. To estimate the time to complete each activity, and to apply these times to the network so that the critical path and overall time of the contract can be determined. Adjustments must be made to the network at this stage if necessary.
3. To determine the resources required and to adjust the network to suit.

POWERS AND DUTIES OF THE RESIDENT ENGINEER

Clause 2 of the ICE *Conditions of Contract* sums up the powers and duties of the resident engineer as:

> to watch and supervise the works and to test and examine any materials to be used or workmanship employed in connection with the works. He shall have no authority to relieve the contractor of any of his duties or obligations under the contract nor, except as expressly provided, to order any work involving delay or any extra payment by the employer nor to make any variation of or in the works.

The resident engineer on a contract is the engineer's representative. His powers are those delegated to him by the engineer (as defined in the *Conditions of Contract*). His powers can therefore *never* exceed those of the engineer.

The powers and duties of the resident engineer should *always* be set out in writing by the engineer at the time of his appointment. This will avoid any misunderstanding due to over-enthusiasm of the RE, as the contractor normally expects to accept the day-to-day instructions from the RE as having the approval of the engineer himself.

The relationship between the resident engineer and the contractor's resident staff was referred to in a report on *The Contract System in Civil Engineering* [33] published by the ICE. That report pointed out that:

> the independence of the engineer and the contractor are unaffected by the existence of a contract, and nothing must be allowed to affect that independence. There is a strong bond of common interest between them, because both wish to see a good construction materialize ... nevertheless each is entitled to his own freedom of thought, different outlook and need of privacy. No matter how small the work may be, no matter how confined the area, the engineer and the contractor should have separate offices. ... The respective staffs serve their head offices and must continually preserve their identity as branch offices.

The duties of the resident engineer and his staff were discussed in a further report by the Institution of Civil Engineers, *The Organization of Civil Engineering Work* [32], and are considered in detail by Twort [76].

An efficient site is usually a happy one, and a spirit of understanding, mutual trust and co-operation between the resident engineer and the contractor (within the limits of both of their responsibilities) can only be to the ultimate advantage of the employer. It is also a fact that the contractor usually obtains no more satisfaction from working with a weak resident engineer than with an over-strict one.

While stressing that none of these detracts from the contractor's responsibilities under the contract, points which should be included in any set of written instructions to a resident engineer are:

1. All work to comply with the specification.
2. Check and agree site levels *before* any bulk excavation is commenced.
3. Check all setting out.
4. Obtain samples of materials.
5. Inspect materials as they are delivered to site.
6. Inspect the formation after excavation.
7. Ensure that adequate timbering of excavations is used.
8. Check reinforcement.
9. Check alignment and level of formwork.
10. Supervise placing of concrete.
11. Organize test cubes of concrete.
12. Inspect concrete on stripping shuttering.
13. Check setting out and levels of sight rails for pipelaying.
14. Check alignment and jointing of pipes.
15. Witness tests on pipelines (details of tests required by the engineer should be clearly stated).
16. Attend routine progress meetings.

17. Keep records of materials and labour employed, if the 'variation of prices' clause is applicable.
18. Order all dayworks in writing, and agree all records daily (N.B.—Authority for any extra works must be given by the engineer.)
19. Keep records of rock excavation.
20. Record all trench depths, etc., for payment purposes.
21. Accurately locate all junctions, etc., on pipelines before they are backfilled.
22. Confirm in writing all verbal instructions to the contractor.
23. Record the state of all buildings (where relevant), fields, gardens, walls, fences, etc., *before* work starts.
24. Provide the engineer with progress reports as required.
25. Maintain a diary of work carried out, the weather, visitors to site, etc.

On overseas contracts the resident engineer may have wider powers delegated to him by the engineer. These may include authority to issue certificates for payment. Where the contract allows the duty-free import of plant and materials, the resident engineer must maintain records of these (supplemented by shipping documents), which must relate details of materials ordered, shipped and incorporated into the works.

SETTING OUT

The responsibility for correct setting out of the works rests with the contractor, unless any error arises due to incorrect information being given by the engineer or the RE. This responsibility of the contractor is in no way relieved by any checking carried out by the engineer or the RE.

Setting out of civil engineering works is referred to by Twort [76] and the survey details are covered by Haywood [72]. A useful leaflet on setting out simple works is published by the Ministry of Public Building and Works (Advisory Leaflet No. 48). While a chain or fibre-glass tape may be adequate for setting out the lines of sewers, manholes, etc., a steel tape should always be used for any setting out where accuracy is a vital factor, e.g. foundations and buildings generally.

For trenches, the centre line and top width should be marked out, and sight rails erected at each manhole position. A third sight rail should be erected along each length as a check against accidental displacement of one rail. Temporary bench-marks should be established at suitable points along the line (clear of the works and all equipment) so that levels can be checked by instrument from time to time.

SAFETY OF EXISTING STRUCTURES

During the excavation for any civil engineering works, particularly trench excavations for sewerage, many other structures become liable to damage. These include buildings, rail tracks, etc., which may be damaged by ground movement; and pipelines, etc., directly damaged during excavation. In this respect, it should be noted that the precautions called for in Clause 22 of the *Conditions of Contract* apply equally to the property of statutory undertakers.

Existing structures can be affected by direct reduction in their support due to excavation in the immediate vicinity, or they may be affected by vibration, by the shrinkage of the soil, or by undermining caused by dewatering operations.

As much information as possible should be collected on other services in the vicinity before excavation begins. These services will include water, gas and electricity, GPO cables, other drains and sewers, and possibly district heating pipes. Details of these *should* be obtainable from the various statutory undertakers or other bodies, but it is often necessary to confirm their location by digging trial holes.

As any pipes, cables or ducts are exposed during excavation, they must be supported to prevent damage and to prevent any interruption in their use. Any damage must be reported immediately to the relevant statutory undertaker, and facilities must be given to them to carry out inspections and repairs. Lamp columns and similar items of street 'furniture' must be protected. Roadside gullies, valve chambers, etc., must be protected from the excavated spoil; it is useful to fill these with straw or similar material before excavation starts, as this will simplify subsequent cleaning.

It is not usually possible to avoid damage to land drains when excavating across fields. Pegs should be driven at ground level to mark the line of these drains as they are found, so that they can be re-laid during the backfilling of the trench. Filling of the trenches to the underside of the land drains must be well consolidated.

MEASUREMENT

In the usual contract, based on a competitive tender, the quantities set out in the bills of quantities are those estimated for the proposed work. The rates in the bills of quantities form the basis of payment to the contractor, but the quantities themselves are re-measured as the work is carried out.

For sewerage work, the variable items are often the depth of trench, class of pipe, type of bedding and type of surface. The ultimate length between manholes will also rarely be the same as the billed length, as manhole positions are frequently adjusted as work progresses.

These 'as constructed' measurements are agreed between the resident engineer and the contractor in accordance with the *Conditions of Contract* (Clause 56), and they form the basis of the contractor's claim for payment. Bulk excavation quantities are usually agreed as the difference between site levels before the work starts and finished levels.

In the *Standard Method of Measurement*, the unit of measurement for sewers and drains is the linear metre, while fittings are measured by number. In practice, it will be found that many engineers adopt a compromise, so that separate items are included in the bills for:

1. Excavation, per linear metre, based on pipe diameter, together with an average depth to invert.
2. Extra over item (1) for the type of surface—in linear metres.
3. Supply, lay, joint and test pipes—in linear metres.
4. Pipe bedding or special protection—per linear metre of trench.
5. Additional items (by number) for junctions and other fittings.
6. Manholes by measured work—separate items for excavation (m^3), concreting (m^3), shuttering (m^2), building in pipes (number), benching (m^3 or m^2), etc.

SITE RECORDS AND DRAWINGS

An important part of the resident engineer's duties is the maintenance of proper records throughout the contract. Drawings must be kept up-to-date to show work actually executed, a day-to-day diary is maintained, progress is recorded on the progress charts, and (most important) records are maintained as the basis of payment to the contractor for work done.

Less obvious items to be recorded and filed for easy reference include:

1. Details of strata, through which trenches, etc., are excavated.
2. Any soil tests carried out.
3. Test results on pipelines.
4. Test results on completed water-retaining structures.
5. Test cubes for concrete.
6. 'Location books' giving accurate details of positions of junctions and other fittings, so that these can be located from reference points on the surface at a later date.
7. Details of buildings, fields, gardens, fencing, etc., before work starts—supported by photographs where applicable.
8. Details of any increase (or decrease) in wage rates or cost of materials. These will not be relevant for the unit rates in the bills of quantities if the variation of price clause is not applicable to the contract, but they will probably affect payments for dayworks.

6 Surface Water Sewerage

SURFACE water sewers provide the means for collection and discharge of natural precipitation, i.e. rainfall, snow, etc. They may either form a part of a 'combined' sewerage system to carry foul domestic wastes in addition to the surface water or they may be 'separate'. In the latter case, a further completely independent system of sewers is provided to deal with the foul sewage. Surface water sewers are usually designed to discharge to the nearest convenient watercourse when they are part of a 'separate' system. If they form part of a 'combined' sewerage system, they will normally discharge to a sewage treatment works; in this case storm sewage overflows may then be provided at suitable points along the sewers so that excess flows can be diverted to neighbouring watercourses in times of storm. As these overflows can be sources of pollution in a stream, the present tendency is towards 'separate' rather than 'combined' sewers, but many existing systems are either combined or at best only partially separate.

The design of a surface water sewerage system begins with a study of the area to be drained. The limits of this area are usually set by natural physical features, and may extend well beyond the limits of any specific development project. The basic requirement is a map of the area to a scale of not less than 1:2500, and preferably to 1:1000, with contours at least at every metre. Existing natural drainage lines can then be marked on the map and the area being considered can be divided into a series of 'drainage areas'. If an Ordnance Survey map is being used, it will normally be necessary to supplement the information provided on the map with an inspection and survey on the ground.

Except for very small areas, the design of surface water sewers is based on rates of rainfall. The rates used will vary in inverse proportion to the size of the drainage area, taking into account the time of concentration and the impermeability of the catchment area.

This chapter summarizes the calculations for surface water sewerage; the subject is dealt with in more depth in the author's book, *Surface Water Sewerage* [65].

RAINFALL

In 1967 the Meteorological Office issued rainfall maps to a scale of 1:625 000 covering the British Isles, illustrating average annual rainfall for the period 1916–50. These maps show that the average yearly precipitation over the whole of the British Isles is of the order of 1100 mm, while the maximum (in parts of North Wales and the English Lake District), is about 5000 mm.

Annual rates of rainfall are, however, only of general interest as far as surface water sewer design is concerned. Much more important are the intensities of rainfall over comparatively short periods of time. The relevant period of time will depend on the 'time of concentration' of the particular drainage area being considered, and will therefore vary with each section of the calculations.

The time of concentration is usually taken as the time taken for the run-off to flow from the point of entry into the sewer or channel, and through the sewer (either as existing or as proposed) to the point where the rate of flow is required. It may then be necessary to add a further period of time to allow for the 'time of entry', when the storm run-off starts some distance away from the point of entry into the sewer. Road Note No. 35 [15] recommended that 'a time of entry of two minutes should be used for normal urban areas, increasing up to four minutes for areas with exceptionally large paved surfaces with slack gradients'.

Various formulae have been put forward for the calculation of rainfall intensities in terms of the duration of the storm. The best-known and most-used in the United Kingdom are probably those which have become known as the 'Ministry of Health' formulae. These take the form:

$$R = \frac{a}{t + b}$$

where
R is the rate of rainfall
t is the duration of the storm
a and b are constants

The original Ministry formulae, expressed approximately in metric units, are:

(*i*) *for* $t = 5$ *to* 20 *min:*

$$R = \frac{750}{t + 10} \text{ mm/h}$$ **Formula 6.1**

(*ii*) *for* $t = 20$ *to* 120 *min:*

$$R = \frac{1000}{t + 20} \text{ mm/h}$$ **Formula 6.2**

where
R is the rate of rainfall in mm/h
t is the duration of the storm in minutes (= time of concentration plus time of entry)

The two Ministry formulae relate to storms of a frequency expected about once per year. In 1935 Bilham published his formula, which related the frequency of storms to their duration and intensity. Expressed in metric terms, Bilham's formula is:

$$T = 1{\cdot}25\, t(0{\cdot}0394\, r + 0{\cdot}1)^{-3{\cdot}55}$$ **Formula 6.3**

where
t is the duration of the storm in hours
r is the total amount of rainfall in millimetres during time t
T is the number of storms of this intensity to be expected in 10 years

It is generally agreed by most engineers that the Bilham formula is cumbersome to use. Recent research by the Meteorological Office has also indicated that it over-estimates those rates of rainfall in excess of 32 mm/h, and the formula has been modified to take this into account. A table of rates of rainfall based on the Meteorological Office formula is published in BSCP 2005. That table gives rates in mm/h, for frequencies from once per year to once in 100 years, and for times of concentration from 4 to 20 min.

The CP also quotes simple 'Ministry type' formulae, based on work originally published by Norris in 1948. The metric equivalents of those formulae, giving rates of rainfall (R) in mm/h, are set out in Table 6.1.

Whichever formula is used, many engineers using traditional methods for calculating sewer sizes have adopted ceiling intensities to avoid over-design. The *Code* quotes ceilings of from 25 mm/h to 38 mm/h. In the examples later in this chapter, a ceiling of 38 mm/h has been adopted.

TABLE 6.1
RAINFALL INTENSITIES

Frequency of recurrence, once in	*Intensity, mm/h*	
	t = 5 to 20 min	*t = 20 to 120 min*
0·5 years	$\frac{580}{t+10}$	$\frac{760}{t+19}$
1·0 years	$\frac{660}{t+8}$	$\frac{1\,000}{t+20}$
2·0 years	$\frac{840}{t+8}$	$\frac{1\,200}{t+19}$
5·0 years	$\frac{1\,220}{t+10}$	$\frac{1\,520}{t+18}$
10·0 years	$\frac{1\,570}{t+12}$	$\frac{2\,000}{t+22}$

The decision as to which frequency of recurrence of storm to adopt in a design is basically one of economics. Sewers designed to carry storms expected, say, once every ten years will be more expensive than those designed for 'one year' storms, but on the other hand the risk of occasional flooding will be less. Practice in the UK is generally to design sewers on storm intensities expected once every year or two years, although the basis of once every five years has been approved in some instances. Whenever possible, the choice of 'design storm' should take into account the cost and effects of any flooding in the locality.

The drainage in the immediate vicinity of buildings should generally be designed on a rainfall intensity of 50 mm/h (see CP 301). This is equivalent to a frequency of about once every two years.

METEOROLOGICAL OFFICE 'STANDARD STORM PROFILES'

Research carried out by the Meteorological Office has suggested that standard storm profiles can be drawn relating the rate of rainfall with the time elapsed since the start of the storm. Separate profiles have been drawn for different frequencies of storm. These are tabulated in Table 8 of Road Note No. 35 [15].

These profiles have been used in the TRRL Hydrograph method of calculating run-off (referred to later in this chapter) and also in a proposed method of analysis of urban rainfall run-off and discharge put forward in a paper by Sarginson and Bourne [59]. That paper provided for the use of the maximum intensity of rainfall from the Meteorological Office profiles, together with allowances for ground slope and storage in the sewers.

The statistics on which the expected mean rates of rainfall and storm profiles are based have been revised as a result of research by the Meteorological Office. For any specified location the

Meteorological Office can supply a table showing the variation of mean rates of heavy rainfall with frequency and duration, i.e. a specially computed equivalent of the Bilham table, and storm profiles that vary with duration. The tables of mean rates of rainfall can be computed to take account of the size of the catchment area.

RUN-OFF

The amount of and rate at which rainfall reaches the sewers is dependent on the extent of the area to be drained, and the relative impermeability of the surfaces over which it has to flow.

An accurate assessment of the area can be made by a survey on site, or with a planimeter on a large-scale map. An approximation is often possible by reference to the grid squares printed on the Ordnance Survey maps. On the larger-scale maps, the lines of the National Grid are at 100-m intervals, so that one square on the map represents $10^4\,m^2$, or 1 ha.

Various formulae and methods have been used in the past for the determination of the rate of run-off. Code of Practice 2005 suggests either the use of the Lloyd-Davies ('Rational') method, or the Transport and Road Research Laboratory Hydrograph method. The Lloyd-Davies method is considered satisfactory for small areas where the maximum sewer size is not likely to exceed 600-mm diameter. The American 'Rational' Formula differs slightly from the Lloyd-Davies Formula in that the impermeability figure is usually taken as varying (like the rainfall intensity) according to the time of concentration.

As an approximate guide to rates of run-off, a rainfall intensity of 50 mm/h (see CP 301) on a 100 % impermeable surface, will produce a run-off of $0{\cdot}14\,m^3/s/ha$ (0·14 cumec/ha).

The Lloyd-Davies formula, which is applied to each section of the catchment in turn, is:

$$Q = 2{\cdot}75 A_p \,.\, R \,.\, 10^{-3}$$ **Formula 6.4**

where

Q is the run-off in cumec
A_p is the area in hectares
R is the rate of rainfall in mm/h

In this formula, the area (A_p) is the 'equivalent impervious area', and is the total area (in hectares) multiplied by an impermeability factor. The choice of a suitable factor depends on a detailed knowledge of the area to be drained, and should always be based on what is to be expected in the foreseeable future, particularly where development is intended. Table 6.2 summarizes various impermeability factors which have been used from time to time.

One of the conclusions reached by Watkins in Road Research Technical Paper No. 55, *The Design of Urban Sewer Systems* [16], is that, 'subject to some qualifications in exceptional circumstances, the whole area of paved surface in an urban area should be considered as impermeable in a sewer design calculation, and the unpaved areas should be considered completely pervious'. This may be rather too general, as the extent of unpaved surfaces in an 'urban area' will vary considerably. Each case should be considered separately, and due allowance must always be made for probable future development.

In the circumstances where the catchment area consists of a large agricultural area discharging to a sewer or stream flowing through a comparatively small built-up area, it may be uneconomical to base the design on the impermeability factors set out in Table 6.2. It is then more satisfactory to

TABLE 6.2
IMPERMEABILITY FACTORS

Type of surface	*Factor %*	*Type of surface*	*Factor %*
Urban areas, where the paved areas are considerable	100	heavy clay soils	70
Other urban areas, average	50–70	average soils	50
residential	30–60	light sandy soils	40
industrial	50–90	vegetation	40
playgrounds, parks, etc.	10–35	steep slopes	100
General development—paved areas	100	Housing development at	
roofs	75–95	10 houses per hectare	18–20
lawns—depending on slope and subsoil	5–35	20 houses per hectare	25–30
		30 houses per hectare	33–45
		50 houses per hectare	50–70

deal separately with the agricultural land and to assess its run-off on the basis of Formula 6.4. The run-off from a rainfall of 40 mm over a period of 24 h is approximately 0·005 cumec/ha.

The TRRL Hydrograph method of design is based on the production of a 'time-area' graph together with a hydrograph of the flow, and has been programmed for calculation by computer. To some extent, it is a development of the earlier Ormsby and Hart method. This method is fully described in Road Note No. 35 and is summarized at Appendix A of BSCP 2005.

In Road Note No. 35 recommendations for the design of carrier or unproductive sewers have been incorporated. These sewers are lengths to which no impermeable area contributes directly. The application to unproductive sewers of some methods of design can result in serious under-design. Procedures to avoid this are given and an explanation of how to apply the TRRL method to overflows and pumping stations has been included.

CALCULATIONS

The following is an extract from calculations for an area to be sewered. It is based on the Lloyd-Davies 'Rational' method, and while the calculations have been set out fully here, they would normally be tabulated to save space, and so that they would be more easily available for future reference. Numbering of the lengths of sewer follows the decimal classification recommended in Road Note No. 35, and refers to the simplified diagram of drainage areas in Fig. 6.1. The hydraulic design of the pipelines themselves is discussed more fully in Chapter 9.

Area No. 1*: Sewer* 3.1

At this stage assume a diameter of 150 mm.
The length is 80 m, and the average surface gradient is 1 in 100.
From Crimp and Bruges tables (see Chapter 9),

Velocity of flow = 0·945 m/s
Time of flow along this section = 85 s
= 1 min 25 s
Add time of entry, say, 3 min 0 s

Total time t = 4 min 25 s

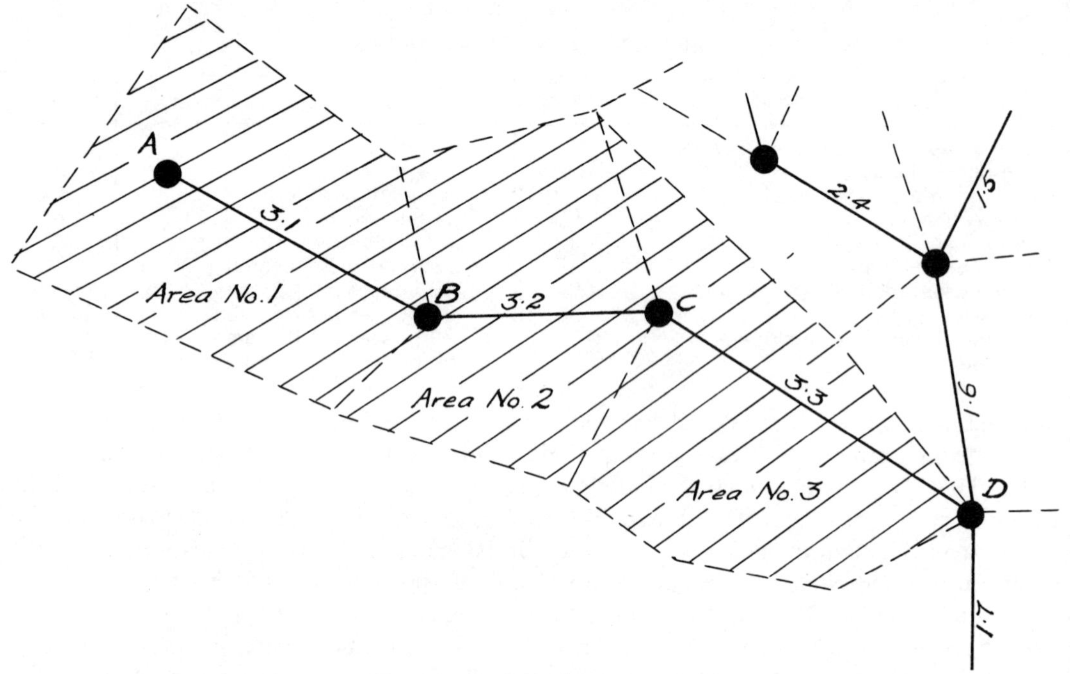

Fig. 6.1. *Example drainage areas.*

This time t is less than 5 min and, as it is not included in Table 6.1, a CP 301 rainfall rate of 50 mm/h can be adopted.

Area No. 1 has an actual area of 0·5 ha; this is now agricultural land, but will be developed in the near future at about 10 houses per hectare.

The equivalent impermeable area of Area No. 1 = 0·5 × 20%
= 0·10 ha

From Formula No. 6.4, run-off = $2{\cdot}75 \times 0{\cdot}10 \times 50 \times 10^{-3}$
= 0·014 cumec

The capacity of a sewer of 150-mm diameter, laid at a gradient of 1 in 100 = 0·017 cumec.

The sewer in the section can therefore be laid to follow the average gradient of the surface.

Area No. 2: *Sewer* 3.2

The average surface gradient in this section is 1 in 145, it may therefore be wise to assume a 225-mm diameter sewer. The length of this sewer will be approximately 70 m.

Velocity of flow	= 1·03 m/s
Time of flow along section	= 68 s
	= 1 min 8 s
Add previous t	4 min 25 s
Total time t for this section	5 min 33 s
	= 5·55 min

Using Table 6.1, for a 'one year' storm,

$$R = \frac{660}{5{\cdot}55 + 8}$$

$$= \frac{660}{13{\cdot}55}$$

$$= 48{\cdot}5 \text{ mm/h}$$

It is, however, proposed to adopt a ceiling of 38 mm/h. Let the extent of Area No. 2 be 0·3 ha; and assume that it is now (and will remain) parkland with a permeability factor of about 25 %. Then the equivalent impermeable area $= 0{\cdot}3 \times 25\%$

$= 0{\cdot}08$ ha (approx.)

Add the area upstream (A_p) 0·10

Total impermeable area to be considered $= 0{\cdot}18$ ha

From Formula No. 6.4, run-off $= 2{\cdot}75 \times 0{\cdot}18 \times 38 \times 10^{-3}$

$= 0{\cdot}019$ cumec

This flow could be carried by a 150-mm pipeline laid at about 1 in 80 or by the 225-mm proposed above. The choice of diameter will depend on the depth of the sewer at manhole 'D', and whether there may be any possibility of future development along section 3.2. A 225-mm diameter pipeline at 1 in 140 will carry 0·043 cumec.

Area No. 3: *Sewer* 3.3

For a length of 102 m and a surface gradient of 1 in 200, assume a diameter of 225 mm.

Velocity of flow is then	0·875 m/s
Time of flow along the section	= 117 s
	= 1 min 57 s
Add previous t	5 min 33 s
Total t for this section	= 7 min 30 s

From Table 6.1, for a 'one year' storm (as before):

$$R = \frac{660}{7{\cdot}5 + 8}$$

$$= \frac{660}{15{\cdot}5}$$

$$= 42{\cdot}5 \text{ mm/h}$$

Again, a ceiling of 38 mm/h will be adopted.

If Area No. 3 is developed with houses at about 30 per hectare, and has an actual area of 0·65 ha,

The impermeable area	= 0·65 × (say) 40 %
	= 0·26 ha
Add the areas upstream (A_p)	= 0·18
Total impermeable area to be considered	= 0·44 ha
From Formula No. 6.4, run-off	= 2·75 × 0·44 × 38 × 10^{-3}
	= 0·046 cumec

If a 225-mm pipeline is to be used, it should therefore be laid at a gradient of 1 in 120 to have a capacity of 0·046 cumec. Alternatively, a 300-mm sewer might be provided; at a gradient of 1 in 200 (to match the surface gradient), its capacity would be 0·077 cumec.

The calculations for each part of the whole area are developed in this manner, from the head of each branch sewer to its junction with a main sewer, and finally along the main sewer to its point of discharge.

STORM SEWAGE OVERFLOWS

Overflows are provided on certain 'combined' sewers so that excess flows at times of storm can be diverted to a convenient watercourse, and the flow in the downstream sewer limited to an agreed maximum figure. A Technical Committee on Storm Overflows and the Disposal of Storm Sewage, appointed by the Minister of Housing and Local Government, issued its final report in 1970. BSCP 2005 recommends that overflows should not be provided on sewers of less than 460-mm diameter, and that the setting of any overflow should, as far as possible, make full use of the available capacity in any sewers downstream. Subject to those considerations, it was usual in the past to set overflows to discharge when the flow upstream exceeded six times the dry weather flow, provided that the watercourse could provide adequate dilution. Based on the recommendations of the Technical Committee, storm sewage overflows should generally be set so that:

$$Q = \text{d.w.f.} + 1360P + 2E$$ **Formula 6.5**

where

Q is the flow passed forward for treatment (in litres/day)
P is the population served by the sewer
E is the volume of industrial effluent (litres/day)

The Committee pointed out that:

'ideally, a storm overflow should achieve the following:

(a) It should not come into operation until the prescribed flow is being passed to treatment.
(b) The flow to treatment should not increase significantly as the amount of overflowed storm sewage increases.
(c) The maximum amount of polluting material should be passed to treatment.
(d) The design should avoid any complication likely to lead to unreliable performance.
(e) The chamber should be so designed as to minimize turbulence and risk of blockage; it should be self-cleansing and require the minimum of attendance and maintenance.'

Fig. 6.2. Automatically raked screens (by courtesy of the Longwood Engineering Co. Ltd).

When existing combined sewers become overloaded due to increases in urban development, it is sometimes more satisfactory to provide overflows on the existing sewers to discharge to a new storm water sewer. This new sewer can then run parallel to the old sewer and would discharge to the sewage treatment works.

An overflow may take the form of one or two side weirs, a stilling pond, or a siphon. In any event, it is essential that overflow starts at the correct rate of flow in the sewer, and that floating solids are retained in the foul sewer as far as possible. Side weirs may be either straight (on one or both sides of the channel) or on the outside of a curve. Weirs should preferably be adjustable, and dip plates must be provided between the line of flow and the weirs to retain solids in the main line of flow. In some circumstances it may be better to provide screens at the overflows; various patent screens are available. One form of automatically raked screen is illustrated in Fig. 6.2.

The Coleman-Smith formula for calculating the length of side weir required is as follows:

$$L = 14{\cdot}9W \,.\, v \,.\, H_1^{0{\cdot}13}\left(\frac{1}{\sqrt{H_2}} - \frac{1}{\sqrt{H_1}}\right)$$ **Formula 6.6**

where

L is the length of weir required in millimetres

W is the mean width of the main channel, or the distance between dip plates in millimetres

v is the mean velocity in the channel in m/s

H_1 is the incoming head above the weir in millimetres

H_2 is the outgoing head above the weir in millimetres

Earlier formulae took into account one value only for the head on the weir (H). The purpose of a well-designed weir is, however, to reduce the outgoing head (H_2) to a minimum and this is often taken as about 20 mm.

It is usually possible to allow for some adjustment of the overflow weir plates in the design, so that the actual overflow level can be altered should the population (and the dry weather flow in the sewer) increase due to development upstream.

OUTFALL STRUCTURES

Where a surface water sewer (or a storm sewage overflow) discharges to a river, the design of the outfall should take into account the need to protect both the stream bed and the banks, the outfall should be compatible with the hydrological characteristics of the river and it should be easily available for inspection and maintenance.

Preferably, the maximum discharge velocity at the outfall should not exceed about 1·5 m/s; if necessary, the energy of the discharging water can be dissipated over a series of steps. A concrete apron and concrete or brick wingwalls should be provided as standard practice. The river bed itself may need protection with an apron of concrete or of gabion mattresses.

BALANCING AND STORAGE

When areas are being developed, it may be found that the existing watercourses will be inadequate for the increased rate of run-off. In some circumstances, some form of balancing or storage of part of the run-off may be preferable as an alternative to the construction of expensive capital works to increase the capacity of a watercourse. Similarly, in some circumstances, it may be found that the provision of a balancing reservoir (or 'water meadow'), together with a smaller-diameter outlet sewer, may be more economical than the construction of a sewer of larger capacity. This is particularly important when an existing sewer through a densely built-up area can be retained, and the cost and inconvenience of duplication can be avoided.

Balancing reservoirs have been used for many years in Europe, and are now being used increasingly in the UK. Considerable impetus to their development has originated in the new and expanding towns and in larger housing schemes on the outskirts of existing towns.

In the author's book *Surface Water Sewerage* [65] he has summarized various graphical methods for the calculation of the storage capacity required. One of the simpler formulae is that prepared by Copas [49] and subsequently simplified by Escritt. If the time of concentration can be neglected the formula is:

$$C = \frac{8{\cdot}02\, A_p^{1{\cdot}5} I^{0{\cdot}5}}{P^{0{\cdot}5}}$$ **Formula 6.7**

where

C is the storage capacity required in m^3
A_p is the impervious area in hectares
I is the number of years between occurrence of storms (see Bilham's formula, Formula 6.3)
P is the rate of outflow from the storage area in cumec

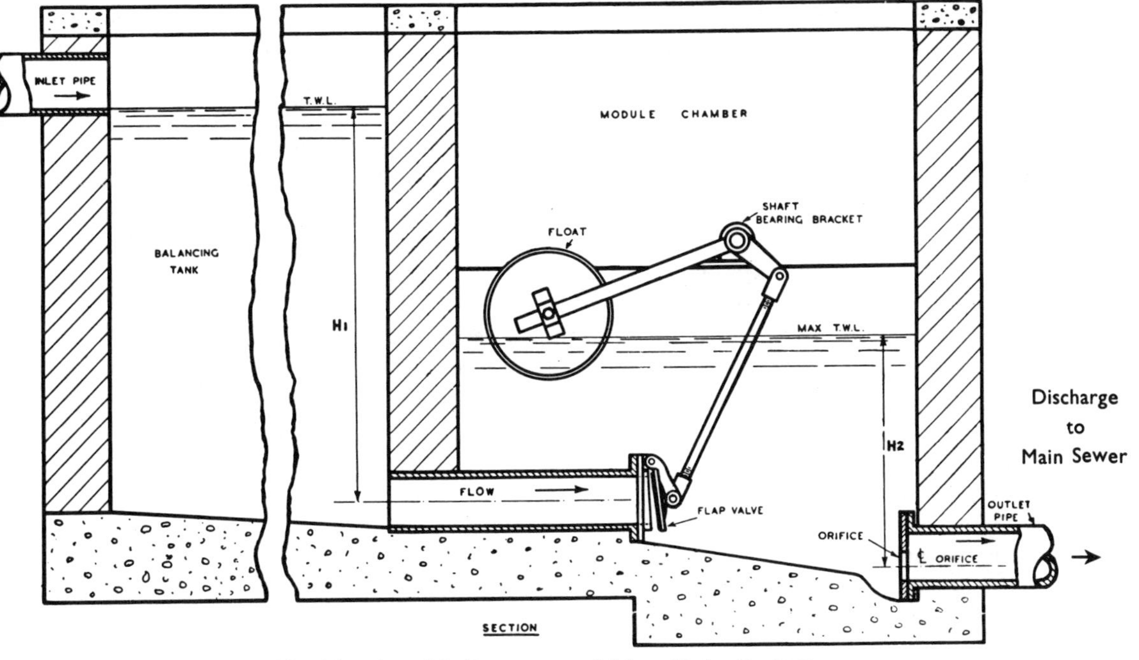

Fig. 6.3. A module (by courtesy of Adams-Hydraulics Ltd).

If the time of concentration is included, two formulae are used, and the solution is based on trial and error to a great extent.

The basic principle of a balancing reservoir is that the rate of outflow is restricted to some agreed maximum figure during normal times, but this may have to be exceeded in times of very severe storm. This restriction can be achieved to varying degrees with a module to regulate the outflow; an automatic electrically operated penstock; or simply by restricting the diameter of the outlet pipe. Figure 6.3 shows a simple module arrangement.

On the normal basis of design, some provision for storage is automatically included in the sewers themselves. In a paper to the Institution of Civil Engineers, Braine [44] suggested that when sewers are designed on the normal 'Ministry' formulae, sewers can be expected to provide for storms which occur once every seven years, if allowance is made for the storage capacity available in a sewer at the beginning of a storm, together with any permissible surcharge, plus the usual practice of providing a pipe size above the theoretical requirements.

SUBSOIL DRAINAGE

Subsoil drainage may be necessary as part of a sewerage system to prevent flooding, or to alleviate or prevent dampness in cellars, etc. The subject is referred to in BSCP 301. The use of mole drainage is considered in more detail in the Ministry of Agriculture, Fisheries and Food 'Growmore Leaflet' No. 44.

Clayware field drains to BS 1196 are manufactured in nominal diameters of 60, 75, 100, 150, 225 and 300 mm. The most commonly used are the 100- and 75-mm diameters. Concrete porous pipes to BS 1194 are available in sizes from 75 to 900 mm. Non-porous socketless vitrified clay pipes with perforations in their lower section are now available in 100-, 150-, 225- and 300-mm diameters.

Subsoil drainage pipes should be laid at depths of 600 to 900 mm, and should generally follow the gradients of the ground surface. They may be set out in various forms (herringbone, fan, gridiron, etc.) to suit local conditions. Herringbone branches should not normally be more than 30 m long, and should be between 6 and 25 m apart, depending on soil conditions—the closer spacing for very stiff clay soils. Outfalls to ditches or other watercourses should be provided as frequently as possible, or if these are not available arrangements should be made for the drainage to discharge through an intercepting trap to a local authority sewer.

The amount of subsoil run-off will be approximately equal to the amount of rainfall which soaks into the ground, but, unlike surface water sewerage design, the maximum discharge occurs long after the passing of the storm. A formula used in America is:

$$R = 116 A \,.\, r \,.\, 10^{-6}$$ **Formula 6.8**

where

R is the run-off in cumec
A is the area of land in hectares
r is the amount of rainfall in mm/24 h

ROAD DRAINAGE

Drainage of road surfaces is a subject in itself, adequately covered by textbooks on road design. This section is intended to provide only sufficient information for the adequate design of drainage from paved areas around pumping stations, at sewage treatment works, and in similar situations.

Average values for crossfall for roads vary from 1 in 35 to 1 in 50, depending on the type of surfacing. Where the longitudinal gradient is flatter than 1 in 200, false gradients of between 1 in 120 and 1 in 200 should be provided between gullies. With a false 'crown' of about 50 mm, gullies would then be spaced at about 25-m intervals.

A general empirical spacing provides gullies at not more than 50 m apart, or one gully for every 200 m^2 of impervious catchment. A formula proposed by Mollinson [56] in a paper to the Institution of Highway Engineers recommended that gully spacing should be as follows:

$$D = \frac{280\sqrt{S}}{W}$$ **Formula 6.9**

where

D is the gully spacing in metres
S is the gradient per cent (for 1 in 25, $S = 4\%$)
W is the width of paved area in metres

Various types of gully pot are available either to BS 539 or 556. For connections to combined sewers, a trapped gully must be used. Gully gratings should be to BS 497; if grey iron castings are used, these should have curved bars when the road gradient is steeper than 1 in 50; the more open ductile iron gratings can be used for any road gradient. Care should be taken in setting the frames of gully gratings with curved bars to ensure that the vertical edges of the curved bars will face upstream when the grating is eventually fixed.

Research into the depth of rainwater flowing over road surfaces has been carried out by the Road Research Laboratory, and this is summarized in their Report LR 236 [18]. The formula in that report can be re-written as:

$$d = 0{\cdot}0474\,(L \times I)^{0{\cdot}5}\,.\,X^{0{\cdot}2}$$ **Formula 6.10**

where

d is the depth of water in millimetres
L is the drainage length in metres
I is the rainfall intensity in mm/h
and the slope is 1 in X

SOAKAWAYS

Wherever possible, surface water drainage from buildings and roads should be discharged into a system of sewers. Occasionally, however, it may be more economical to discharge this to soakaways. Individual soakaways may be provided at each gully, etc., or a number of gullies may be connected to a sewer (normally 150-mm diameter) which itself discharges to a soakaway.

A soakaway may be a pit provided with a roof slab and with open-jointed (i.e. pervious) base and sides. Frequently the pit is filled with rubble; while the rubble serves no real purpose and in fact takes up part of the capacity of the soakaway, it may obviate the need for a roof slab designed to take traffic loading. The rubble should, however, be sealed at the top with some form of concrete or stone cover to prevent the ingress of soil from above. The capacity of a pit without rubble can be calculated accurately, but this is not possible when rubble is used. Small precast 'domestic' soakaways have capacities of up to 0·65 m^3 per metre depth, while specially manufactured precast concrete segmental soakaways are obtainable, with capacities of up to 3·5 m^3 per metre depth.

The usual basis of design for the capacity of a soakaway is to allow a storage volume sufficient for a minimum of 10 mm of rainfall over the area to be drained, based on 100% impermeability. As a formula (based on a rainfall of 10 mm):

$$Q = 0{\cdot}01A$$ **Formula 6.11**

where

Q is the capacity in cubic metres (m^3)
A is the area to be drained, in square metres (m^2)

It should be remembered that soakaways are only satisfactory in permeable soils or where they can be excavated down to a permeable stratum. They must also discharge above the maximum subsoil water level.

CONDITIONS OVERSEAS

Formulae used for the estimation of rainfall and run-off in the UK will not necessarily apply to any overseas scheme. Unless formulae have already been prepared in a country, it may be necessary to

TABLE 6.3
OVERSEAS RAINFALL (DURATION OF 'ONE YEAR' STORMS OF 50 mm/h)

Location	*Time in minutes*
Africa: Nairobi	16
Burma: Mandalay	30
Burma: Rangoon	64
Thailand: Bangkok	45
Ceylon: Colombo	100
United Kingdom	5

analyse rainfall records so that curves can be plotted for various types of storm. If possible, these records should extend over two or three decades at least. Data for any less period must be used with caution.

The normal standard rainfall gauge is not satisfactory for this work, and one must have access to records either from 24-hour recording type rain gauges (pluviographs) such as the Dines tilting siphon autographic rain gauge or from special rainfall intensity gauges. An intensity gauge can be used to record directly the hydrographs of storms, plotting the rainfall intensity in mm/h against the time.

When using charts from an autographic gauge the amount of rainfall during any storm of a certain duration must first be plotted against the number of occurrences of such a storm during any one year's records. If this is plotted on 'log-log' graph paper, 'one year' storm intensities can then be interpolated for each period of time. If these storm intensities are then plotted (to a natural scale) against the duration of storm (to a log scale), it is possible to draw a reasonable straight-line curve so that a formula for rainfali intensity in terms of duration can be prepared after a little trial and error. This formula can usually take the form of:

$$R = \frac{a}{t + b}$$

where R and t have the same meanings as in Formulae 6.1 and 6.2, and where a and b are constants.

Table 6.3 compares the approximate durations of 'one year' storms at 50 mm/h intensities based on formulae prepared from the rainfall records of a number of overseas countries. Based on the information available a few years ago, a formula for a 'one year' storm for Rangoon (Burma) was:

$$R = \frac{7000}{t + 70} \text{ mm/h}$$ **Formula 6.12**

Statistics have been published for many years by the United States Weather Bureau [99] and from these local formulae have been calculated. Figures suggested for the Atlantic seaboard cities of USA give a formula of:

$$R = \frac{3000}{t + 20} \text{ mm/h}$$ **Formula 6.13**

Storm intensity figures for Australia have been published in the Report of the Stormwater Standards Committee of the Institution of Engineers, Australia [79] and formulae for 'one year' storms are given for different localities.

7 Foul Sewerage

WHEREAS the design of surface water sewers is dependent on rainfall intensities and the extent of the catchment area, rates of flow of foul sewage are dependent on the distribution of the population and on the rate at which water is used. The average flow is called the 'dry weather flow' (d.w.f.) and is the average rate of flow of domestic and industrial wastes, together with any infiltration, measured after a period of seven consecutive days of dry weather during which the rainfall has not exceeded 0·25 mm.

The pattern of the discharge of water after use is very similar to that of water demand, and in consequence it is general to assume that the *peak* flow of domestic sewage (excluding industrial wastes), is the same as that of water demand, i.e. about twice the average rate of flow (2 d.w.f.). When the sewers are completely 'separate' they should be designed to carry up to four times the dry flow (4 d.w.f.) to allow for weekly and seasonal peaks. Where infiltration is to be expected, or where the sewerage system is not strictly separate, a maximum flow of 6 d.w.f. is generally assumed to make some allowance for the surface water.

The figure for dry weather flow to be used for design purposes is generally not available when a scheme is being drawn up, and it is usually calculated from the population to be served in the future, multiplied by an allowance *per capita* (in terms of either litres or cubic metres per day). Additional capacity must then be included for the waste waters from industrial premises, farms, etc., over and above this calculated domestic flow. When further development upstream of the proposed sewer is possible, a suitable allowance for this should be added when deciding on the capacity of the new sewer.

When old sewers or house drains are to be connected to a new system of sewers, it is generally accepted that additional provision must be made in the new sewers for infiltration of surface and subsoil water into the older sections. The amount of this allowance should be based on actual site investigations as far as possible.

THE SEPARATE SYSTEM

While most new sewerage projects are now designed as 'separate' systems, many older schemes may be either 'combined' or 'partially separate'. BSCP 2005 points out that, subject to the various considerations of design and cost, 'where local conditions permit, the present-day tendency is to favour the adoption of the separate system'.

In a separate system, surface water is excluded from any sewer which carries foul sewage. Two completely separate sewerage systems are then required, the first to carry toilet, bath and all other domestic wastes, industrial wastes, etc., and the second for roof and yard drainage, road drainage and other surface water. When sewers are designed to be 'separate', *all* surface water, *including* that from back roofs and yard gullies, must be rigidly excluded from the foul sewers.

The construction cost of two completely separate systems of sewers is, of course, generally higher than that of a combined system, and there may be some problems in finding suitable points for the discharge of the surface water. A separate system, however, has the advantages that the foul sewers, being of smaller diameter, can be designed to be self-cleansing; and that any pumping stations and the sewage treatment works can generally be of a more economical design.

COMBINED AND PARTIALLY SEPARATE SYSTEMS

When a 'combined' system is used, only one sewer is then provided in a road to take both foul and surface waters. The capacity of this sewer will be set principally by the amount of surface water to be carried, and the capital cost is therefore usually less than that of two separate systems. With only one sewer there is also no risk of a wrong connection. Storm sewage overflows are normally necessary to restrict the flow during times of peak rainfall run-off, and storm sewage tanks must be provided at the sewage treatment works.

The design of storm sewage overflows is referred to in Chapter 6. These overflows are always a weak point on any combined system, as they can be responsible for the pollution of watercourses, and are liable to cause complaint on grounds of amenity.

The 'partially separate' system is often a necessary compromise between the separate and combined systems. When existing sewers and house drains already carry some surface water it may be more economical to design any new sewerage system on the assumption that some surface water from roofs and yard gullies will be allowed into the foul sewers. This will reduce to some extent the cost and the problems of the drainage of surface water from buildings, but the overall cost will be more than for a combined scheme. The design of sewers, pumping stations and treatment works is made more complicated, as it may be difficult to assess the actual amount of surface water which will discharge to the sewers. With a 'partially separate' system, a full survey of all properties is necessary before any design for extension of sewers or sewage treatment is commenced.

POPULATION FIGURES

Foul sewer design is dependent on the distribution of population, and the amount of water used *per capita*. In view of the long length of life of sewers and similar works, it is essential that any design is based on the probable *future* population densities and not on existing populations, unless there is a tendency for the population to decrease. It is generally wise to work on an estimated population for the next twenty-five to thirty years.

Where relevant, the official statistics of the Registrar-General can be used as a guide to existing population densities, together with the local registers of voters, bearing in mind that these only apply to persons over eighteen years of age and do not include the names of aliens. The local planning authority can advise on possible future densities of population and probable location of new industries. At holiday resorts and similar areas, peak population densities in the holiday season are more important than the resident population density.

Increases in population due to recent building or future development will generally be available in terms of the number of new houses or in terms of the proposed number of houses per hectare. An allowance of 3·0 to 3·5 persons per dwelling is now a fairly common minimum for sewer design purposes.

RATES OF FLOW

Water usage *per capita* tends to increase continually, and is now rarely less than 140 litres *per capita* per day. Code of Practice 2005 suggest that a *maximum* domestic sewage dry weather flow could be taken as 230 litres *per capita* per day, but higher figures have been used from time to time, especially in the design of sewers for new towns where every house has modern plumbing facilities.

At 3·5 persons per house, the figure of 230 litres is equivalent to a domestic dry weather flow of about 800 litres per dwelling per day. For an absolutely separate sewerage design, using peak flows of 4 d.w.f., this would give daily flows of 920 litres per head or 3200 litres per dwelling. Corresponding figures on the basis of 6 d.w.f. would be 1380 and 4800 litres per day respectively. These figures, along with their approximate equivalents in terms of cubic metres per second, are tabulated in Table 7.1.

When the available information is in terms of population densities, and assuming the same figure of 230 litres *per capita* per day, the dry weather flow can be calculated from the following formulae:

$$Q = \frac{D \times A}{375} \text{ l/s} \qquad \textbf{Formula 7.1}$$

or

$$Q = \frac{D \times A \times 10^{-3}}{375} \text{ cumec} \qquad \textbf{Formula 7.2}$$

where

Q is the dry weather flow in either l/s or cumec
D is the population density expressed as persons per hectare
A is the area in hectares

It is normal to add an allowance for infiltration of subsoil water into the sewers. The actual amount will depend on the age of the sewers and house connections, the type of pipe joints used, and the height of the water table in the subsoil. This is referred to in more detail in Chapter 13. A figure of 12 litres per hour per 100 m of sewer has been suggested by Bevan and Rees; this would seem to be a reasonable figure for new sewers constructed with flexible/mechanical joints. Higher figures (from 40 to 100 litres per hour per 100 m) were often used in the past. When new foul sewers are designed to carry a maximum flow of 6 d.w.f. it can be assumed that this figure *includes* an allowance for infiltration.

TABLE 7.1
UNIT DISCHARGE RATES (BASED ON A DRY WEATHER FLOW OF 230 LITRES PER CAPITA *PER DAY)*

Unit	*Litres/day*	*Cumec*
Per capita d.w.f.	230	$2{\cdot}66 \times 10^{-6}$
4 d.w.f.	920	$10{\cdot}6 \times 10^{-6}$
6 d.w.f.	1 380	$16{\cdot}0 \times 10^{-6}$
Per dwelling d.w.f.	800	$9{\cdot}25 \times 10^{-6}$
4 d.w.f.	3 200	$37{\cdot}0 \times 10^{-6}$
6 d.w.f.	4 800	$55{\cdot}5 \times 10^{-6}$

N.B. 10^6 l/day = 0·016 6 cumec; 1 cumec (m^3/s) = 90×10^6 l/day (approx.).

INDUSTRIAL AND OTHER FLOWS

In addition to the flows calculated from the normal residential population figures, further allowances must be made for the flows from schools, hospitals, factories, etc. General practice is to assume a daily flow of from 70 to 100 litres *per capita* from day schools. For boarding schools, hospitals and similar establishments, the flow can be taken as the same as that from normal dwellings, care being taken to base this on the maximum possible number of residents.

The allowance to be made for the domestic wastes from industrial premises should be based on a detailed survey, as the average and peak rates of discharge will be dependent on the availability of cooking facilities, canteen, etc. It is usual to allow for at least 50 % of the rate of flow from normal domestic premises, i.e. if the dry weather flow from domestic premises has been taken as 230 litres per head per day, the minimum allowance for domestic sewage from industrial premises should be 115 litres for each person employed at the factory.

The rate of discharge of *industrial* waste water will vary from factory to factory and will be very dependent on the type of processes employed. Whenever possible, the allowance for the maximum daily rate of flow from an industrial establishment should be based on actual measurements or records. When this is not possible (e.g. when designing sewerage for a new industrial estate), an allowance can usually be made on the basis of experience elsewhere in the locality or according to the types of industry to be expected. Average figures for the quantities of waste water from various industries have been published from time to time by the Water Research Centre; certain industries are referred to by Isaac [73] and in a paper to the Institution of Public Health Engineers published in January 1969. The volume of industrial waste water can sometimes be very considerable, and may form a substantial part of the total flow in a sewerage system.

RURAL SCHEMES

The basis of design for rural schemes, in addition to population trends, must also take into account the discharges from farms and other rural industries, and the possible changes in methods of farming. The domestic *per capita* water consumption is, however, generally lower, and, as compared with a figure of 230 for urban populations, a figure of 150 litres *per capita* per day for rural sewerage schemes would probably be conservative.

Comparative rural domestic flow figures per person would then be:

	Litres/day	*Cumec*
4 d.w.f.	600	$7{\cdot}0 \times 10^{-6}$
6 d.w.f.	900	$10{\cdot}5 \times 10^{-6}$

Additional allowances for the wastes from farm animals will depend on the type of animals kept and the method of farming employed. For a normal dairy herd, to allow for washing down and milk cooling, a figure of 100 litres per day should be included for each cow. A further figure of 50 litres per head per day should be added for other cattle where the drainage from cattle sheds will connect to the sewer. The Water Research Centre [22] has reported figures of up to 20 litres per day for pigs when these are kept for fattening on larger farms.

The inclusion of any industrial waste waters will usually have a very marked effect on the design flows in rural schemes, and each case must be investigated in detail. Some flows (per tonne of vegetables treated) from typical rural industries quoted by the Water Research Centre are:

Vegetable washing	. .	from 2000 to 7000 litres
Vegetable canning	. .	4500 to 30 000 litres
Vegetable freezing or dehydration	. .	13 000 to 30 000 litres

PETROL AND OIL TRAPS

Where a garage wash-down or similar installation could give rise to the possibility of petrol or oil being washed into a drain or sewer, a petrol interceptor must be fitted. A suitable design based on the requirements of the GLC and other local authorities is shown in Fig. 7.1. For single private

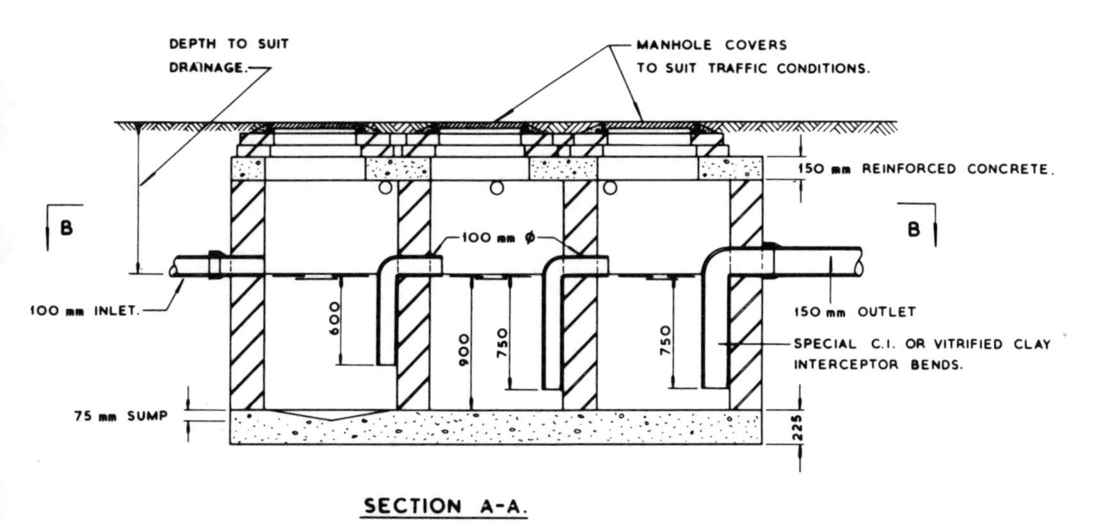

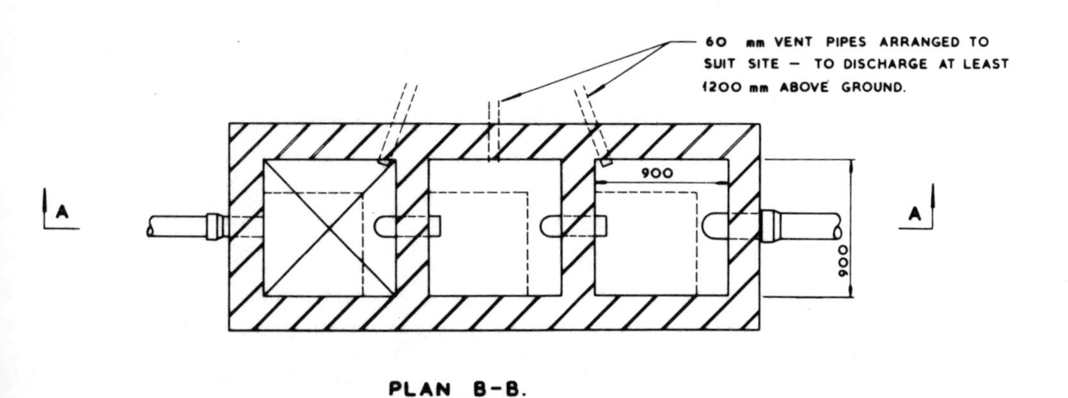

Fig. 7.1. Petrol interceptor.

garages, a deep gully trap similar to Table 24b of BS 539 or to Fig. 57 or 58 of BS 1130 may be sufficient. It should be provided with a perforated lifting tray so that debris and grit can be removed.

Oil skimmers are available (either fixed or floating pattern) for the recovery of floating oil from sumps or lagoons. These consist of revolving cylinders which collect the oil and discharge it to a pump sump. Depending on viscosity and layer thickness, oil can be removed at rates of about 1 litre/min for each 100 mm of revolving cylinder.

A literature survey on *Oil Interceptors for Surface Drains* prepared by the Hydraulics Research Station in 1976 [31] lists thirty-five sources of information published in the UK and overseas.

8 House Drainage and Small Schemes

THE design of house drainage and smaller sewerage schemes presents certain problems which do not occur with the larger municipal schemes. Rates of flow are smaller and flow is more intermittent; more junctions and short pipe lengths are required.

The subject of 'Building Drainage' is covered by CP 301, which looks at both foul and surface water drainage, including the drainage of properties where a public sewer is not available.

The site information needed for a scheme of building drainage will include details of any available public sewers (location, diameter, depth, etc.), any easements required to enable connections to be made and any specification requirements by the local authority (basis of design, materials of construction).

An application under the Town Planning Acts and the Building Regulations will include details of drainage proposals and points of connection to the public sewers. CP 301 recommends that existing sewers should be shown in black on drawings, proposed foul drains in red, proposed surface water drains in blue, and any trade effluent drainage in yellow.

FOUL DRAINAGE

The peak flow in a drain can be assessed from details of the number of appliances to be connected and their likely peak frequency of use. Reference can be made to CP 304 'Sanitary Pipework above Ground' for information on typical peak rates of discharge.

It is usual to assign to each sanitary appliance a 'discharge unit value' or 'fixture unit value'. These are quoted in CP 304 and by Wise in his book covering the hydraulic design of pipework in dwellings [77]. Hartley, in a paper to the Institution of Public Health Engineers [53], plotted the flow in a drain against the number of discharge units. The rates of flow quoted in that paper are in cubic feet per minute; these can be converted to cumec (m^3/s) with no appreciable error by dividing them by 2000. Tables in CP 304 give the maximum number of discharge units to be allowed with various diameters of vertical stacks and horizontal branches.

The design of foul drains and sewers for smaller schemes is usually based on six times the average dry weather flow (6 d.w.f.). At a d.w.f. of 230 litres/person/day, this peak flow per person will then be 0·016 litres/s. Assuming an occupancy rate of about 3 persons per dwelling, this latter figure can then be changed to 0·05 litres/s *per dwelling*, giving a simple multiplier to use when designing drainage schemes for new developments.

For industrial waste discharge, the peak rates of flow must be assessed on the water requirements of the particular industry, bearing in mind the hours of working and the probable peak discharges from any manufacturing process.

SURFACE WATER

Where the main surface water drain on a site is not more than about 180 m in length, or when the area to be drained is less than two ha, unless there is exceptional risk of damage due to flooding, it is usual to use a design rainfall intensity of 50 mm/h, irrespective of the theoretical time of concentration (see Chapter 6). Experience has shown that such a provision is likely to prevent flooding more frequently than once in 20 years. However, where the roof of a factory (or similar premises) discharges via valley gutters and internal downpipes, it may be wiser to allow up to 75 mm/h rainfall in the design of the drains immediately downstream, to avoid any backing up in the downpipes and consequent flooding at the valley gutter.

The estimate of run-off will normally be based on a 100 % flow from all impermeable areas (roofs and paved areas) including 50 % of the exposed vertical faces of tower buildings (see CP 301). This will give a run-off of 14 litres/s for each 1000 m^2 of impermeable area.

Where possible, surface water drains should be connected either to a public sewer or to a suitable ditch or stream. In some circumstances (and provided the subsoil conditions are suitable) it may be preferable to construct soakaways. These should be sited well away from building foundations and should have a minimum storage capacity of 13 m^3 for every 1000 m^2 of impermeable area; this is equivalent to a storage space for 13 mm of rainfall.

LAYOUT OF DRAINS

The layout of the drains around houses and other buildings will be dictated, to a great extent, by the positions of downpipes and yard gullies. It is normal practice, and required by the Building Regulations, to provide an inspection chamber at each point where there is a change of direction or gradient. A chamber is also usually provided at the head of each drain run, but for smaller drains a rodding eye may sometimes suffice.

The Building Regulations require an inspection chamber within 12·5 m of any junction of one drain with another, unless there is an inspection chamber or manhole at the junction. On a long length of drain no part of that drain should be more than 45 m from a chamber, i.e. no drain (between two inspection chambers) should be more than 90 m in length.

According to the definitions incorporated in the Building Regulations, an inspection chamber 'means a chamber constructed on a drain so as to provide access thereto for inspection and cleansing'. All inspection chambers sited *within buildings* should be absolutely watertight.

The overall layout should be as simple and as direct as possible, making the best use of the natural slopes of the ground to keep the depth of excavation (and therefore the cost) to a minimum. Branches should join another drain at an angle *with the flow*, or with a swept 90° junction, and *never* at 90° to or against the flow.

Special care is needed when designing small drainage schemes to ensure that the surface water and foul drains have sufficient *vertical separation*, so that drains can cross over each other if necessary and so that branches from either side can be connected without difficulty. Allowing for two drain runs, each 100-mm diameter, there should be a vertical clearance from *invert to invert* of not less than 200 mm; this allows for one pipe diameter, plus pipe thicknesses and socket depths. If the lower drain is of a larger diameter, the vertical clearance must be increased accordingly.

The subject of drainage works *above* ground is adequately covered by CP 304, by various Ministry of Public Building and Works leaflets and by the book by Wise of the Building Research Establishment [77].

HYDRAULIC DESIGN

In the past, considerable use was made of McGuire's Rule for the drainage in the immediate vicinity of buildings. This laid down minimum gradients of 1 in 40 for 100-mm, 1 in 60 for 150-mm and 1 in 90 for 225-mm diameter pipelines. The rigid adoption of this rule without consideration of other factors has often resulted in bad drainage design, where the upper sections have been extremely shallow, where unduly steep gradients have resulted in uneconomic depths of drains or where the pipes have been of a larger diameter than necessary.

Building Research Station Digest No. 6 Second Series [7] stresses that a 100-mm-diameter drain laid at a gradient of 1 in 70 can adequately serve as many as twenty houses. Provided the flow is sufficient to give a self-cleansing velocity once per day, a 150-mm pipeline can be laid to a gradient as flat as 1 in 150 and at that gradient could be expected to serve up to 100 houses. The practice of using a 150-mm diameter drain instead of a 100-mm diameter drain merely so that a flatter gradient can be theoretically adopted is most unsatisfactory, as it reduces the depth of liquid flowing in the pipeline, with the result that a self-cleansing velocity of flow may never be attained and the possibility of silting and blockage is therefore increased. The increase in the cost of the larger diameter pipes is rarely offset by any savings in excavation costs.

CP 301, 'Building Drainage', suggests that at the head of a drain serving only a few houses, a 100-mm pipeline should in general not be laid flatter than 1 in 40. Where a drain is within about 12 m of a w.c., this gradient may not, however, be necessary; research on this by the Building Research Establishment is referred to in Chapter 9. The minimum diameter of pipe permitted for domestic foul wastes by the Building Regulations is 100 mm. Although 75-mm diameter is permitted for surface water drains in those Regulations, it is better practice to adopt a minimum diameter of 100 mm for both foul and surface water drains (see CP 301).

For a foul drain, the size of the pipe and the gradient at which it is to be laid should be chosen so that, where there is a danger that traps will be siphoned, the depth of flow does not exceed 75 % of the pipe diameter. At that depth of flow the capacity of the pipeline is about 90 % of its full velocity.

CP 301 points out that a satisfactory gradient therefore depends upon the peak flow as follows:

1. For peak flows of more than 1·0 litres/s but less than 2·5 litres/s, the gradient should be not flatter than 1 in 70 for a 100-mm drain. In general this gradient is satisfactory if the drain serves the equivalent of at least one w.c.
2. For peak flows of 2·5 litres/s or more, the gradients should not be flatter than 1 in 130 for a 100-mm drain or 1 in 200 for a 150-mm drain.

As the gradients quoted depend on high standards of workmanship, it is recommended that, in practice, 100-mm pipelines should not be flatter than 1 in 80, and 150-mm pipelines not flatter than 1 in 150. Where flows are likely to be very small, steeper gradients may be required.

MATERIALS

The normal pipes used for small drainage schemes are of vitrified clay, uPVC or pitch-fibre. Occasionally iron or asbestos-cement pipes will be used where drains pass under buildings.

The Building Regulations requirements for a drain or sewer which passes through a building are that it shall:

a. be adequately supported throughout its length without restricting thermal movement, any fitting giving such support being securely attached to the building, and
b. be so placed as to be reasonably accessible throughout its length for maintenance and repair.

Vitrified clay pipes should be to BS 65 and 540. They are often of the plain-ended type with plastic sleeve couplings as these can be cut to length as required and jointed with a standard coupling. Plain-ended pipes are available in 100- and 150-mm diameters. Various special adapters are available to allow for easy connections from soil and downpipes, w.c. outlets and waste pipes. Alternatively, spigot and socket pipes with the 'O'-ring flexible joints should be used. Rigid joints of cement mortar are now rarely used either for sewers or for house drains. A wide range of special fittings is available, including bends, junctions, saddles and rest bends.

uPVC pipes for drainage are available in 110- and 160-mm nominal diameters to BS 4660. Again a wide range of fittings is available and the pipes can be easily cut to any required length, making this type of pipe eminently suitable for house drainage. Care must, however, be exercised in using a smoke test with uPVC pipelines as some smoke generating devices produce smoke which is detrimental to plastic pipework.

Pitch-fibre pipes (to BS 2760) are available in diameters ranging from 75 to 200 mm, complete with snap ring-type joints and a full range of fittings.

PUMPING

While pumping on a small scheme should be avoided if possible, there will be occasions when the relative levels of the site and the existing sewers make the pumping of foul sewage unavoidable. Surface water run-off from a small development should *never* be pumped as the wide variations in flow would make it both uneconomical and unreliable.

Whenever possible a pumping station should be designed and located so that it will be adopted by the local authority, who will then maintain it. The site should not be subject to flooding and it should be located as far as possible from residential property. Some provision should be made for an emergency overflow to operate in the event of mechanical or power failure; the overflow being where it will cause least disturbance or damage to property.

Generally, areas of new development will be drained on the separate system and the sewage flow from small individual areas may be too small to suit the output of standard unchokable centrifugal pumps. In these cases, the adoption of ejectors or special small submersible pumps will be necessary, the pumps being capable of dealing with flows up to 6 d.w.f.

The main problem associated with pumping small flows is the necessity of maintaining a self-cleansing velocity in the rising main, and yet, at the same time, pumping the sewage sufficiently frequently to avoid any problems with septicity. Where the sewage is neither screened nor macerated, the usual minimum pump capacity must be about 20 m^3/h to give a self-cleansing velocity in a 100-mm diameter main. It will be found, however, that few manufacturers are able to market a pump as 'unchokable' and suitable for solids up to 100 mm, with an output of less than 50 to 60 m^3/h.

These limitations can be overcome to some extent by the use of ejectors or by using special types of pumps which either divert the solids around the impeller, or which incorporate some form of macerator. Small pumps of this type are available with outputs as low as 2 m^3/h and for use with a 30-mm diameter rising main. This subject is covered more fully in the author's *Pumping Stations for Water and Sewage* [64].

9 Hydraulic Design

USING the information obtained from the earlier surveys, and working on a map of 1:2500 or 1:10 000 scale, as is relevant, it is now possible to set out lines of possible sewers for either the surface water run-off or for foul sewerage. The possible points of discharge (e.g. watercourses for surface water sewers, and treatment works for foul sewers) will set the overall pattern of the system of sewers. The sewers themselves should then follow the fall of the surface of the ground as far as possible, as this will reduce depths of excavation or, alternatively, will reduce pumping costs to a minimum. Pumping stations and rising mains are considered in Chapters 14 and 15.

Having adopted a possible network of main sewers, sections along these can be drawn, together with any possible alternatives, so that the surface profiles can be studied and trial pipeline gradients compared with the surface gradients to obtain the most economical trench depths.

Depending on whether the system is to be separate or combined, the quantity of sewage to be carried by each length of sewer (in m^3/s) can now be calculated from the formulae in Chapters 6 and 7. The flow which any pipeline can carry depends on its cross-section (and therefore on its diameter), its gradient or slope in relation to the horizontal and a friction coefficient. This coefficient will depend on the material of the pipes and the liquid being carried. The diameter of pipeline and the required gradient for any specific flow of sewage can be found from an empirical formula or from tables.

When using either formulae or tables, it should be remembered that sewerage includes a number of 'unknowns'. As population and run-off figures are often based on intelligent guesses of possible future developments, there is no logic in aiming at extreme accuracy in any calculations using those figures. It is therefore general to err slightly on the side of over-design, but one must not be tempted to use this as an excuse for careless hydraulic design.

Every sewer and drain should be designed to have a self-cleansing velocity of flow of 0·75 m/s at least once per day. The velocity of flow in a circular pipe is the same when it is running half-full as when it is full; velocities and discharges at various depths of flow can be obtained from Table 9.1. In the past, maximum flow velocities were often limited to about 3·0 m/s to reduce the effect of scour from grit carried in the sewage. This is not now considered to be important, except in some special circumstances, such as at bends in large-diameter pipelines. The Working Party Report on Sewers and Water-Mains has drawn attention to the economies in capital cost that can be achieved by permitting higher velocities, the saving being largely through the elimination of backdrop manholes.

FORMULAE AND TABLES

Of the many formulae produced in the past for the calculation of capacity and velocity of flow in pipelines, probably the two in more general use for sewer design in the United Kingdom are those of Crimp and Bruges, and Manning. Other formulae include those by Colebrook-White, Hazen-Williams, Kutter, Chezy, Bazin and Darcy. The Hazen-Williams formula has been used in the chapter on rising main design (Chapter 15). The Colebrook-White formula was used by the

TABLE 9.1
PIPES RUNNING PARTLY FULL—PROPORTIONATE VALUES OF VELOCITY AND DISCHARGE

Prop. depth	*Prop. velocity*	*Prop. discharge*	*Prop. depth*	*Prop. velocity*	*Prop. discharge*
0·01	0·0890	0·0002	0·51	1·0084	0·5170
0·02	0·1408	0·0007	0·52	1·0165	0·5340
0·03	0·1839	0·0016	0·53	1·0243	0·5513
0·04	0·2221	0·0030	0·54	1·0319	0·5685
0·05	0·2569	0·0048	0·55	1·0393	0·5857
0·06	0·2892	0·0071	0·56	1·0464	0·6030
0·07	0·3194	0·0098	0·57	1·0533	0·6202
0·08	0·3481	0·0130	0·58	1·0599	0·6374
0·09	0·3752	0·0167	0·59	1·0663	0·6546
0·10	0·4012	0·0209	0·60	1·0724	0·6718
0·11	0·4260	0·0255	0·61	1·0783	0·6889
0·12	0·4500	0·0306	0·62	1·0839	0·7060
0·13	0·4730	0·0361	0·63	1·0893	0·7229
0·14	0·4953	0·0421	0·64	1·0944	0·7397
0·15	0·5168	0·0486	0·65	1 0993	0·7564
0·16	0·5376	0·0555	0·66	1·1039	0·7730
0·17	0·5578	0·0629	0·67	1·1083	0·7893
0·18	0·5775	0·0707	0·68	1·1124	0·8055
0·19	0·5965	0·0789	0·69	1·1162	0·8215
0·20	0·6151	0·0876	0·70	1·1198	0·8372
0·21	0·6331	0·0966	0·71	1·1231	0·8527
0·22	0·6507	0·1062	0·72	1·1261	0·8680
0·23	0·6678	0·1160	0·73	1·1288	0·8829
0·24	0·6844	0·1263	0·74	1·1313	0·8976
0·25	0·7007	0·1370	0·75	1·1335	0·9119
0·26	0·7165	0·1480	0·76	1·1353	0·9258
0·27	0·7320	0·1594	0·77	1·1369	0·9394
0·28	0·7470	0·1712	0·78	1·1382	0·9524
0·29	0·7618	0·1834	0·79	1·1391	0·9652
0·30	0·7761	0·1958	0·80	1·1397	0·9775
0·31	0·7901	0·2086	0·81	1·1400	0·9892
0·32	0·8038	0·2217	0·82	1·1399	1·0004
0·33	0·8172	0·2352	0·83	1·1395	1·0110
0·34	0·8302	0·2489	0·84	1·1387	1·0211
0·35	0·8430	0·2629	0·85	1·1374	1·0304
0·36	0·8554	0·2772	0·86	1·1358	1·0391
0·37	0·8675	0·2918	0·87	1·1337	1·0471
0·38	0·8794	0·3066	0·88	1·1311	1·0542
0·39	0·8909	0·3217	0·89	1·1280	1·0605
0·40	0·9022	0·3370	0·90	1·1243	1·0658
0·41	0·9132	0·3525	0·91	1·1200	1·0701
0·42	0·9239	0·3682	0·92	1·1150	1·0732
0·43	0·9343	0·3841	0·93	1·1093	1·0752
0·44	0·9445	0·4003	0·94	1·1027	1·0757
0·45	0·9544	0·4165	0·95	1·0950	1·0745
0·46	0·9640	0·4330	0·96	1·0859	1·0714
0·47	0·9734	0·4495	0·97	1·0751	1·0657
0·48	0·9825	0·4662	0·98	1·0618	1·0567
0·49	0·9914	0·4831	0·99	1·0437	1·0419
0·50	1·0000	0·5000	1·00	1·0000	1·0000

Hydraulics Research Station in their Paper No. 4 [30]. Escritt has published a set of 'Sewer and Water-main Design Tables' in imperial and metric units [70].

The formulae and tables prepared by Crimp and Bruges [68] over eighty years ago are simple to use in comparison with many other formulae and are used by many engineers. The formula for velocity of flow in pipes or channels, modified for use with metric units is:

$$v = 0{\cdot}834\sqrt[3]{r^2}\sqrt{s}$$ **Formula 9.1**

where

v is the velocity of flow in m/s
r is the hydraulic mean depth in millimetres
s is the hydraulic gradient, i.e. the fall divided by the length

For pipes running full or half-full, the equivalent formula is then:

$$v = 0{\cdot}33\, d^{2/3} I^{-1/2}$$ **Formula 9.2**

where

v is the velocity in m/s
d is the diameter in millimetres
I is the inclination (length divided by the fall)

Based on Formula 9.2, the discharge, when running full, can then be found from the following formula;

$$Q = 26 \times 10^{-8} d^{8/3} I^{-1/2}$$ **Formula 9.3**

where

Q is the discharge in cumec
d is the diameter in millimetres
I is the inclination

The tables and diagrams by Crimp and Bruges which were published originally in 1897, are based on these formulae and are widely used for the designs of sewers. Earlier calculations based on imperial units can be converted to *approximate* metric units as follows:

i. $$\frac{\text{Velocity in ft/min}}{200} = \text{velocity in m/s}$$

ii. $$\frac{\text{Discharge in cumins}}{2000} = \text{discharge in cumec}$$

The Manning formula is:

$$v = \frac{0{\cdot}01}{n} r^{2/3} s^{1/2}$$ **Formula 9.4**

where

v is the velocity of flow in m/s
n is a coefficient of friction
r is the hydraulic mean depth in millimetres
s is the hydraulic gradient

Using the same terminology as for the Crimp and Bruges formula above, the Manning formula for pipes running full or half-full is then:

$$v = \frac{0{\cdot}004}{n} d^{2/3} I^{-1/2}$$ **Formula 9.5**

It will be seen that Formula 9.2 is the same as this formula, when the coefficient of friction (n) is taken as 0·012. The Crimp and Bruges formulae and tables have this value of the coefficient built into them.

Gravity pipelines are rarely designed to run full. The calculated daily peak flow in the sewer must therefore be expressed as a proportion of the total capacity of the pipeline as calculated from Formula 9.3. From Table 9.1 it is then possible to obtain the proportionate depth of flow, and the proportionate velocity at that depth. That velocity should not be less than 0·75 m/s.

FRICTION COEFFICIENTS

Since the publication of a number of empirical formulae late in the nineteenth century, many tables of proposed friction coefficients have been prepared. These are probably more relevant in connection with the design of larger-diameter pipelines and channels, particularly those constructed to carry clean water. They are also important in the design of inverted siphons and rising mains (see Chapter 15).

For gravity sewer design, the following quotation from CP 2005 is relevant:

> It needs to be borne in mind that the flow capacity of a sewer after a short period of use may depend upon the characteristics of the slime growing or deposited on the pipe wall and upon the deposit of grit on the invert. For this reason it is no longer regarded as necessarily axiomatic that sewers of up to 900-mm diameter be designed for hydraulic roughness dependent only upon the material of which the pipes are made. The actual roughness to be taken will depend upon the quality of workmanship, the accuracy with which joints are centred, and the build-up of slime or grit which the designer may think reasonable before the sewer requires maintenance.

Experiments on small-diameter pipes of different materials carrying foul sewage have indicated that any advantages of initial smoothness are quickly lost due to the build-up of slime; also that this sliming and the consequent increased roughness values of mature pipes are substantially independent of the nature of the pipe material. Hydraulics Research Paper 4 [30], puts forward suitable values of k_s for use with the Colebrook-White formula as 0·6 mm, 1·5 mm, and 3·0 mm, according to whether the sewer condition was 'good', 'normal' or 'poor' respectively. A paper by Ackers [41] suggested that those values of k_s in the Colebrook-White formula made suitable allowance for the build-up of slime in drains or sewers constructed of *any* material. A k_s value of 1·5 mm over the lower 25 % of the pipe surface is more or less equivalent to a coefficient of $n = 0{\cdot}012$ in the Manning formula.

The values given for velocities and discharges in the Crimp and Bruges tables (being based on a coefficient of $n = 0{\cdot}012$) therefore satisfactorily take into account the effects of the build-up of slime and grit in the sewers and drains after a short period of use. It follows that those tables can

TABLE 9.2
FRICTION COEFFICIENTS FOR USE WITH MANNING'S FORMULA

Material	*Coefficient*
Cast-iron pipes	0·013 to 0·017
Concrete	0·015
Well-planed timber	0·008
Good brickwork or stonework	0·015
Old brickwork or stonework	0·020
Corrugated metal culverts	0·021
Trimmed earth, canal banks, etc.	0·025 to 0·030

therefore be used for pipes of any materials. For larger pipelines (over about 900-mm diameter) and for culverts, the values of n to be used with Manning's formula can be taken from Table 9.2.

SEWER GRADIENTS

Gradients in sewers and drains should be such that at least once each day the velocity of flow will be self-cleansing. In a pipeline which is not surcharged, the hydraulic gradient will normally be more or less parallel with the line of the pipes. It is general practice, therefore, to design the gradient of the invert of a drain or sewer and to assume that it will not flow surcharged.

On the same basis, the soffit of a pipeline will represent the hydraulic gradient when the pipe is running full, but not surcharged. For this reason, the lines of soffits of sewers should be continuous, and any steps due to increase in diameter of the pipes should be formed in the invert. In practice this means that the invert of a 150-mm pipeline discharging from a manhole should be set 50 mm lower than the invert of any 100-mm pipeline discharging to the manhole. This will ensure that the sewers and drains are properly ventilated, as a continuous soffit line will allow all gases to travel upwards towards the head of the system under any conditions of flow.

The Building Research Establishment has carried out some research into the effects of very flat gradients in house drainage. Pilot experiments have shown that within about 12 m of a w.c. flush, the hydraulics of a drain are dominated by the flush wave, and the gradient of the drain does not seriously influence the frequency of blockage. The Building Research Establishment has reported one such 100-mm diameter drain laid at 1 in 1200 with no history of blockages. It is, however, stressed that a very flat gradient is not satisfactory if the drain is long or if it is badly laid.

TYPICAL SEWER CALCULATIONS

Example 1. A surface water sewer, designed on a 'one year' storm basis, must be capable of carrying 0·4 cumec. The gradient of the surface of the ground is about 1 in 400. Sewers discharging into this sewer will be up to 450-mm diameter.

From Crimp and Bruges tables, the full capacity of a 600-m sewer laid at a gradient of 1 in 400 would be 0·347 cumec. This is less than the required capacity of 0·4 cumec.

The next larger diameter (675 mm) at the same gradient of 1 in 400 has a full capacity of 0·476 cumec.

This larger diameter would be satisfactory, as the velocity of flow when full would be 1·29 m/s.

At 0·4 cumec, the proportionate discharge would be 0·4/0·476 = 0·84 and the proportionate velocity (see Table 9.1) would be 1·12. The velocity of flow at 0·4 cumec is therefore 1·12 × 1·29 = 1·45 m/s.

It is possible, however, that site conditions may be such that a saving could be made by using the smaller sewer (600-mm diameter) at a steeper gradient. It will be seen from Crimp and Bruges tables that to obtain a *full* capacity of 0·4 cumec, the required gradient for a 600-mm sewer is 1 in 300. At that gradient the full velocity would be 1·37 m/s.

At least two solutions are therefore possible:

i. A 675-mm sewer at 1 in 400

or

ii. A 600-mm sewer at 1 in 300

Example 2. A 'partially separate' sewer connection is required to drain 225 houses. An existing 300-mm diameter sewer is available with an invert level of 100·00 at about 480 m from the site boundary. The lowest ground level of the site is at this boundary and is 106·50.

From Chapter 7 it will be seen that 6 × d.w.f. from 225 houses at a discharge rate of 230 litres per person per day, will be:

$$225 \times 55{\cdot}5 \times 10^{-6}\,\text{cumec} = 0{\cdot}0125\,\text{cumec}$$

Reference to Crimp and Bruges tables shows that a 150-mm pipeline laid at 1 in 140 will carry 0·0146 cumec.

The full velocity in this pipeline at 1 in 140 would be 0·799 m/s.

At 6 *d.w.f.* (*the maximum flow*)

The proportionate discharge would be $\dfrac{0{\cdot}0125}{0{\cdot}0146} = 0{\cdot}855$

The proportionate depth would be 0·71

The proportionate velocity would be 1·1231 × 0·799 = 0·9 m/s

At 2 *d.w.f.* (*the normal daily peak rate of flow*)

Discharge rate = 2/6 × 0·0125 cumec
= 0·0042 cumec

The proportionate discharge would be 0·27

The proportionate depth would then be 0·36

The proportionate velocity would be 0·8554 × 0·799 = 0·68 m/s

The gradient must therefore be increased to give a minimum velocity of 0·75 m/s at least once per day.

Try a gradient of 1 in 100:

The capacity of a 150-mm pipeline at 1 in 100, when full, is 0·0172 cumec

Proportionate discharge at 2 d.w.f. $= \dfrac{0{\cdot}0042}{0{\cdot}0172} = 0{\cdot}24$

Proportionate depth is then 0·33

Proportionate velocity is then 0·81

Full velocity at 1 in 100 is 0·945 m/s

Velocity at 2 d.w.f. is therefore $0{\cdot}81 \times 0{\cdot}945 = 0{\cdot}765$ m/s

A satisfactory connection would therefore be a 150-mm diameter drain at a gradient of 1 in 100. The fall required in 480 m of pipeline is then 4·8 m.

The connection would join the main sewer with level soffits, i.e. the invert should be 150 mm above that of the existing 300-mm sewer. The new sewer will therefore have a discharge invert level of $100{\cdot}00 + 00{\cdot}15 = 100{\cdot}15$ and an invert at the housing site end of $100{\cdot}15 + 4{\cdot}80 = 104{\cdot}95$. The depth of the sewer at the boundary of the site would then be $106{\cdot}50 - 104{\cdot}95 = 1{\cdot}55$ m.

INVERTED SIPHONS

An inverted siphon is used when a gravity sewer must cross a valley as an alternative to following the normal gradient of the sewer, which would entail constructing it above ground level. There is always the possibility of silting in inverted siphons, but the deposition of grit is less likely in a separate system than in a combined or partially separate system.

An inverted siphon is a pipe (or a series of pipes in parallel) connecting from the lower end of one gravity sewer, dropping under a valley and rising again at the other side to connect to the head of another gravity sewer. There must be sufficient difference in the levels of the two sewers to give a suitable hydraulic gradient in the siphon.

Section 6.3 of CP 2005 refers to the design of inverted siphons, and points out that as the rates of flow in a separate system will vary from 0·33 to 4·0 d.w.f., it will seldom be possible to arrange for a single pipe siphon to be self-cleansing and that consequently multiple-pipe siphons are normally advisable.

The velocity at minimum flow (0·33 d.w.f. for separate sewers) should not be less than about 1·2 m/s and the forebay of the siphon should be arranged so that the various pipes of the multiple-pipe siphon can come into use consecutively. All pipes do not need to be of the same diameter, and often two pipes only are sufficient. Flushing arrangements can sometimes be arranged if suitable self-cleansing velocities cannot otherwise be obtained.

Calculations for the head loss in an inverted siphon should be based on the losses due to friction through the pipe or pipes, *plus* entry losses, ,*plus* losses due to the bends. The friction losses can be obtained from Crimp and Bruges tables, but some engineers prefer to use one of the empirical formulae, with a suitable friction coefficient based on the material of the pipeline. Those formulae are considered in Chapter 15.

Entry losses can be calculated from the following formula:

$$H = \frac{kv^2}{2g}\ \text{m} \qquad \textbf{Formula 9.6}$$

where

v is the velocity in m/s
g is the acceleration due to gravity = 9·806 m/s^2
k is a constant

In Formula 9.6 the value of k is usually taken as 0·5, the entry losses are then $0{\cdot}025v^2$ m (approx.). The losses due to each bend in the pipeline can also be calculated from the same formula:

$$H = \frac{kv^2}{2g}\,\text{m}$$

where v and g have the same meaning as before.

The constant k can be taken as 0·5 for normal medium CI bends, and 0·75 for short bends.

The total losses due to entry and bends (but *excluding* friction losses) in a normal inverted siphon using four medium bends would then be:

$$H = 5 \times \frac{0{\cdot}5v^2}{2g} \quad \text{or} \quad 0{\cdot}125v^2\,\text{m (approx.).}$$ **Formula 9.7**

where v is the velocity of flow in m/s.

The total losses calculated from Formula 9.7 above must then be added to the friction losses in the pipeline, which are, of course, dependent on the length of the siphon.

10 Structural Design

RESEARCH into the loads on buried pipelines was begun in Iowa, USA, in 1908, and various reports on the 'Marston Theory' were published from 1913 onwards [55]. This work was summarized by Clarke and Young in a paper to the Institution of Civil Engineers in 1959 [46] and enlarged upon in later papers [47, 48]. The whole subject of structural design has been summarized and presented in concise form in a book by Clarke entitled *Buried Pipelines: A Manual of Structural Design and Installation* [67].

Part A of the normal Water Authority estimate form relates to 'sewer pipes with flexible joints which, together with their bedding, are designed to carry specific loads'. That part of the form assumes either that complete computations of the total external design loads have been made, or that use has been made of approved tables. Pipes under roads or verges should normally be laid with at least 1·20 m of cover, and under fields and gardens with at least 0·90 m of cover.

Part B of the form can be used as an alternative to Part A. This retains the original 'Ministry of Health requirements' for pipes *not* designed to carry specific loads. These can be summarized as follows:

1. Pipes of 750-mm diameter and under, with 6·00 metres or more of cover in trenches, to be surrounded with at least 150 mm of concrete. Pipes of more than 750-mm diameter under the above conditions may require additional concrete protection.
2. Subject to (1), all pipes with over 4·3 m of cover to be bedded on and haunched with at least 150 mm of concrete to at least the horizontal diameter of the pipe. The splaying of the concrete above that level to be tangential to the pipe.
3. Subject to (1), all pipes of 450-mm diameter and over to be bedded and haunched with at least 150 mm of concrete to at least the horizontal diameter of the pipe. The splaying of the concrete above that level to be tangential to the pipe.
4. Subject to (5), all pipes under 300-mm diameter and with less than 4·3 m of cover may be laid without concrete, if the joints are of the socket or collar type.
5. All pipes with less than 1·20 m of cover under roads (except roads formed of concrete) or 0·90 m not under roads to be surrounded with at least 150 mm of concrete.
6. In all cases the filling or concrete support must be well rammed and consolidated at the sides and haunches of the pipe. Selected filling should be used up to at least 300 mm above the top of the pipe.

Many new types of joint have been introduced by pipe-manufacturers, and most pipes can now be obtained with flexible joints to satisfy Part A of the form. It is therefore to be expected that very few designs will be based on the old 'Ministry requirements'.

A buried pipeline, complete with its bedding, should be considered as a structure, and the onus is on the designer to calculate the type of bedding required, with due regard to the crushing strength of the pipes to be used, the position of the pipeline in relation to ground levels, the type of subsoil, any superimposed traffic loadings, and the width of the trench to be excavated. It should, however, be borne in mind that correct design must be supported by a good specification and correct methods of construction. The use of the best methods of design will not compensate for bad construction, and vice versa.

In 1962, the Building Research Station issued National Building Studies Special Report No. 32, *Simplified Tables of External Loads on Buried Pipelines* [1]. To obtain simplicity, those tables were based on a combination of probable 'worst' loading conditions. Special Report No. 37 [3] was subsequently issued in 1966 so that a more accurate and economical design could be produced. In January 1967 the Road Research Laboratory published results of research into the effects of impact of moving vehicles [19, 52].

A revised edition of *Simplified Tables of External Loads on Buried Pipelines* [12] was published by the Building Research Station in 1970 to take into account the recommendations of the Working Party Second Report, and to include both imperial and metric units. These revised tables include three basic tables (for the three types of traffic loading, see below), and a fourth to show the breakdown of the figures. The tables are based on one soil density (2000 kg/m^3) and one soil friction coefficient for narrow-trench conditions (0·130). The value of $r_{sd}p$ has been taken as 0·5 generally, with 0·7 for pipes of 300-mm diameter and under.

The Ministry of Housing and Local Government Working Party on the design and Construction of Underground Pipe Sewers issued its Second Report in October 1967. That report took into account the research of the Building Research Station and the Road Research Laboratory, and recommended, *inter alia*, that a rational design of pipe sewers based on the Marston theory should be accepted.

The Second Report put forward a set of design criteria to enable pipelines to be designed 'on a safe but not extravagant basis'. They can be summarized as follows:

1. Use of the Marston theory as set out in National Building Studies Special Report No. 37, with certain modifications.
2. Design for superimposed loads from traffic as follows:
   (*a*) In main traffic routes and under roads which are liable to be used for temporary diversions of heavy traffic, provision to be made for British Standard No. 153 type HB loading, with an impact factor of 1·3.
   (*b*) In other roads, except access roads used only for very light traffic, provision should be made for a maximum of two wheel loads, each of 7250 kg static weight, spread 0·90 m apart, acting simultaneously, with an impact factor of 1·5.
   (*c*) For sewers laid in fields, gardens and lightly trafficked access roads, provision should be made for a maximum of two wheel loads, each of 3200 kg static weight, spread 0·90 m apart, acting simultaneously, with an impact factor of 2·0.
3. Special provision for distributed surcharge loads should not normally be necessary.
4. A 'safety margin' of 80 % of the guaranteed minimum crushing strength should be adopted with unreinforced pipes. The 'minimum proof test strength' to be used for reinforced concrete pipes.

THE MARSTON THEORY

The early work by the late Professor Marston, as developed by Schlick and Spangler, is set out fully in a manual issued by the Water Pollution Control Federation of America [84] and in the textbook by Clarke [67]. The reader requiring full information should consult one of those books. This chapter sets out a summary of the Marston theory for normal usage.

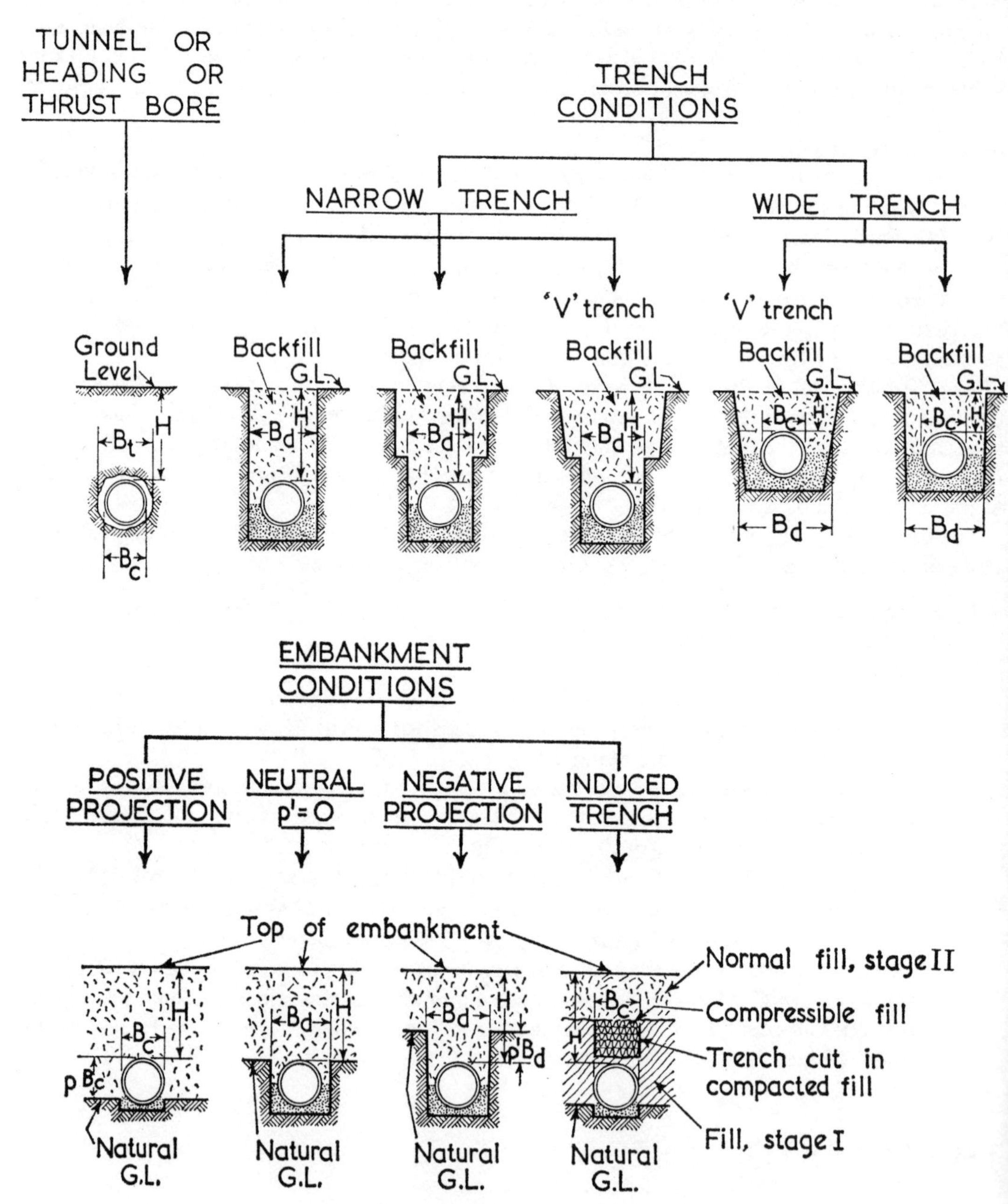

Fig. 10.1. Classification of fill loading conditions imposed by various construction methods (from Buried Pipelines, by courtesy of N. W. B. Clarke).

The three basic classifications for the determination of loads on buried pipelines are:

1. Trench conditions
2. Embankment conditions
3. Headings and tunnels

The first two conditions are generally referred to in the United Kingdom as 'narrow-trench' and 'wide-trench' respectively. The loads on pipelines in heading and tunnel can be determined by using the formula for narrow-trench conditions, with suitable modification to allow for cohesion.

In narrow-trench calculations, the resultant soil load on the pipeline is taken as the difference between the weight of the prism of soil within the trench (and above the top of the pipe), and the friction or shearing force between the prism of soil and the sides of the trench. Under wide-trench conditions, those shearing forces can act downwards under certain conditions, to increase the effective loading.

Experiments carried out in America support Marston's theory of the effects of the friction planes. The classifications of fill loading conditions imposed by various methods of construction are illustrated in Fig. 10.1.

PIPE BEDDING

In 1958 [45] Clarke published details of research into types and causes of fractures of brittle pipes. He stressed the importance of using flexible-telescopic joints, and the need for a correct and uniform bedding. The types of bedding proposed by Marston and Spangler were illustrated in NBS Special Report No. 32. The more important of these are shown at Fig. 10.2, while the bedding factors normally used in design calculations are given in Table 10.1.

Specifications for the granular bedding material and concrete were referred to in Chapter 4. For smaller diameter pipes (up to about 300 mm), it is normal to use granular material which will pass a 19-mm sieve, but be retained on a 4·75-mm sieve. The Clay Pipe Development Association has

TABLE 10.1
BEDDING FACTORS

Bedding factor, F_m	*Bedding class*	*General description*
1·10	Class 'D'	Hand-trimmed trench bottom
1·50	—	Thrustbores
1·90	Class 'B'	Granular bedding and back-filling to mid-diameter
2·20	Modified Class 'B' (Class 'S')	Granular surround to a minimum of 100 mm above the pipe soffit
2·60	Class 'A'	Unreinforced concrete cradle (120°C) or arch (180°)
3·40	Class 'A_{rc}'	Reinforced concrete cradle (120°) or arch (180°) with reinforcement of 0·4% of the concrete area
4·80	Class 'A_{rc}'	Reinforced concrete cradle or arch with reinforcement of 1·0% of the concrete area

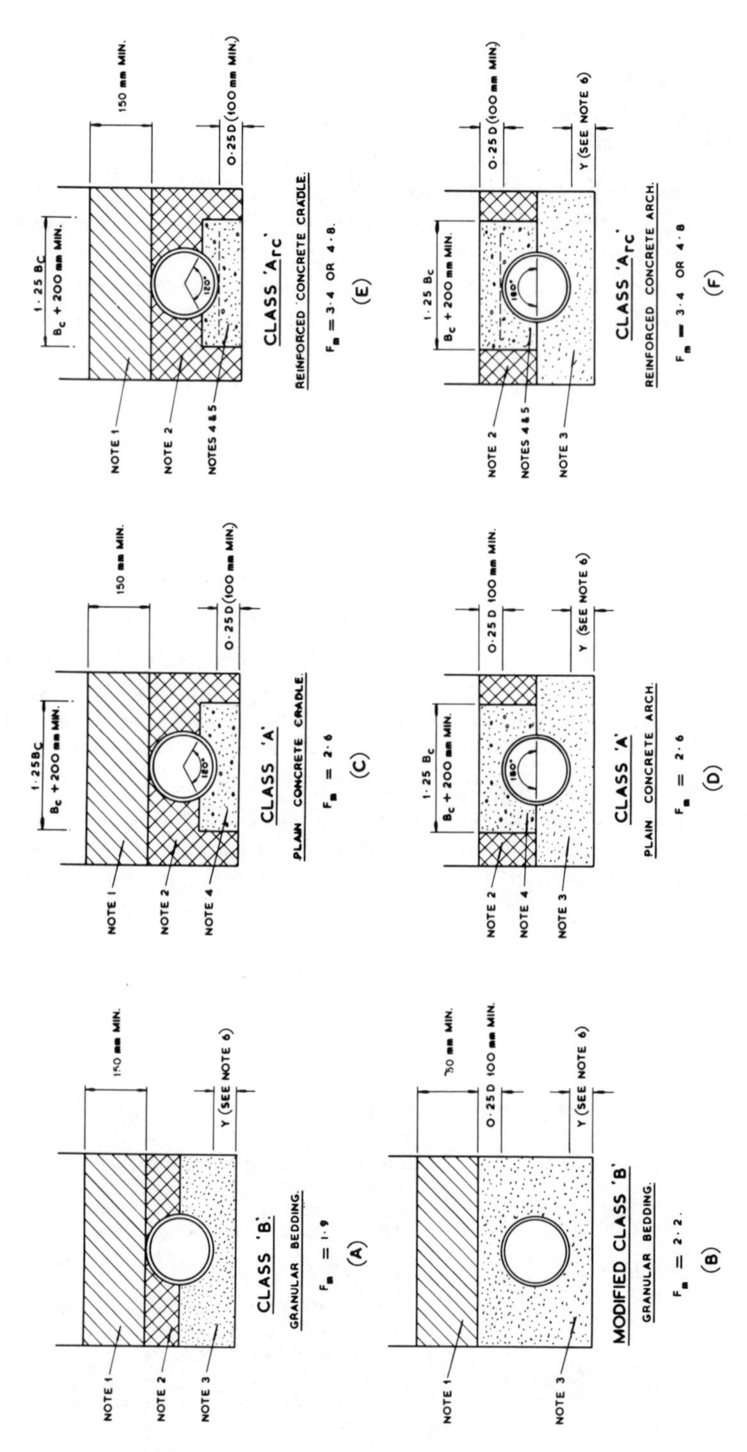

Fig. 10.2. *Types of pipe beddings (see p.* 89 *for Notes).*

recommended the use of material approximating to Table 1 of BS 882, 'Aggregates for Concrete', and has proposed that the nominal size for 100-mm pipes should be 10 mm, for 150-mm pipes 10 or 14 mm and for 225-mm and larger pipes 10, 14 or 20 mm. Not less than 85 % should pass a 13·2-mm sieve (ISO Series R40/3) for pipes up to 300-mm diameter, and not less than 85 % pass a 19-mm sieve for pipes over that diameter. Comparison of the two ISO series of sieves and the original BS series (BS 410) is given in Fig. 4.1 (Chapter 4).

In January 1966 Storey published suggestions for granular bedding materials for larger diameter pipes [61]. His general conclusions were that for 1800-mm diameter pipes, on a specific scheme at Ipswich, a good free-draining compactible material had the following properties:

ISO sieve size	*Per cent passing*
26·5 mm	100
9·5 mm	35–75
500 μm	10–25[a]
125 μm	0–5

[a] This applied to 1050-mm pipes only.

Early practice was to use single-sized pea gravel for bedding pipes, but this is not now recommended. It is more expensive than graded gravel or broken stone, and sharp angular particles are more stable and reduce the tendency of a pipe to settle into a bed. An ideal bedding material should provide uniform support, should not compact or migrate under load, and should be available at low cost.

Various government departments have issued recommended specifications for granular material suitable for bedding and sidefill of flexible pipelines (PVC and pitch-fibre). In general, the granular material should not be larger than about 20 mm, but particles up to 40 mm may be allowed provided that the amount passing a 19-mm sieve shall not exceed 5 % of the total weight. Granular bedding for flexible pipes should have a minimum depth of 100 mm below the pipes. Sidefill should be of the same material, firmly compacted in 80-mm layers between the pipe and the sides of the trench and continued over the crown of the pipe for a depth of 100 mm (for 100-mm diameter pipes), or 150 mm for larger pipes.

Notes to Fig. 10.2.

1. Filling material to be free from lumps, stones and roots; lightly compacted by hand.
2. Filling material to be free from lumps, stones and roots; carefully compacted around the pipes.
3. Granular bedding material.
4. Concrete with a minimum strength of $2{\cdot}1 \times 10^6$ kgf/m^2 (20·6 MN/m^2) at 28 days.
5. 0·4 % reinforcement for a bedding factor of 3·4
   1·0 % reinforcement for a bedding factor of 4·8
6. (a) In rock or mixed soils:
   $Y = 0{\cdot}25B_c$ under barrels, with a minimum of 200 mm under both barrels and sockets.
   (b) In machine-dug uniform soils:
   $Y = 1/6B_c$, with a minimum of 100 mm under both barrels and sockets.
   (c) In hand-shaped uniform soils:
   $Y =$ 100 mm minimum under both barrels and sockets.

CONCRETE PROTECTION

The traditional methods of providing concrete protection were with a concrete bed, a bed and haunch, or a bed and surround. All of these are retained in BSCP 2005, 'Sewerage'. The proposals put forward by Marston and Spangler for concrete cradle and concrete arch have been referred to earlier in this chapter and are illustrated in Fig. 10.2.

The bedding factors quoted in Table 10.1 for concrete protection are based on the use of concrete with cube strengths of $1{\cdot}4 \times 10^6$ kgf/m^2 at the time of loading with backfill, and $2{\cdot}1 \times 10^6$ kgf/m^2 after 28 days. These two figures are approximately equal to 13·75 MN/m^2 and 20·6 MN/m^2 respectively.

The National Clay Pipe Institute of America has proposed [80] the use of a bedding factor of 4·5 for a full surround of unreinforced concrete, with concrete of the same strength as that specified above. The dimensions recommended for a surround to give this bedding factor are set out in Table 10.2.

TABLE 10.2
UNREINFORCED CONCRETE SURROUND—BEDDING FACTOR 4·5

Nominal diam. of pipe mm	*Dimensions of concrete surround*	
	Under pipes and at sides mm	*Over pipes mm*
150	100	100
300	100	100
375	100	100
450	125	125
525	125	125
600	150	150
675	175	150
750	200	150
825	200	150
900	225	150
975	225	150
1 050	250	150

With acknowledgements to the National Clay Pipe Institute, USA.

The NBS Special Report No. 38, *High-strength Beddings for Unreinforced Concrete and Clayware Pipes* [4] includes charts of bedding factors for 180° reinforced concrete cradles and arches for various steel contents from 0·3 % to 1·0 %. These indicate that bedding factors as high as 9·0 are possible with correctly designed beddings. These high-strength beddings are particularly suitable for larger diameter pipes, and it is suggested that for clayware pipes the values of F_m greater than 5·0 should be taken as tentative at present.

The recommended concrete protections include both cradle and arch. For the same loading, the increase in supporting strength provided by a 180° arch is the same as that provided by a 120° cradle, while the arch has a number of advantages on site. When arch protection is to be provided, the pipes can be laid as a continuous operation on a granular bed, and jointing and testing can be completed before any concrete is placed. The concrete itself can be put into position more satisfactorily without horizontal construction joints and without the possibility of contamination from the muddy trench bottom. It must, however, be borne in mind that in certain circumstances the effective loading on the pipeline *may* be greater with an arch (as the width of the arch is used in lieu of the pipe diameter in calculations). This is referred to later in this chapter.

RIGID PIPES IN TRENCHES

In the following formulae and calculations the notation of earlier publications has been retained as far as possible, to avoid confusion. Units of measurement have however been revised as set out in Table 10.3. As pipe strengths have been quoted in kgf/m in various BS specifications, the unit of pressure in these calculations has been taken as the kgf/m; 1·0 kg will produce a downward force of

TABLE 10.3
NOTATION AND UNITS USED

Notation	*Description*	*Unit of measurement*
B_c	Outside diameter of pipe or width across concrete arch	mm
B_d	Effective width of trench	mm
B_t	Effective width of heading	mm
D	Internal diameter of pipe (nominal)	mm
H	Height of cover over top of pipe	mm
H_i	Height of cover over invert of pipe	mm
W_c	Fill load—narrow trench condition	kgf/m
W_c'	Fill load—wide trench condition and positive projection	kgf/m
W_{csu}	Equivalent distributed load from a concentrated surcharge load	kgf/m
W_w	Load caused by water contained in the pipe	kgf/m
W_e	Total effective external distributed load	kgf/m
W_T	Laboratory strength of pipe, i.e. 'safe crushing test strength'	kgf/m
W_f	Field strength of bedded pipe when buried	kgf/m
γ	Density of soil	kg/m³
C_d	Load coefficient in narrow trench	—
C_c	Load coefficient in wide trench	—
C_t	Load coefficient for tunnel and heading	—
F_m	Bedding factor	—
F_s	Safety margin	—
F_i	Impact factor for live loads	—
K	Rankine's coefficient of internal earth friction	—
μ	Coefficient of internal friction in the fill = tan ϕ	—
μ'	Coefficient of friction between fill and trench sides = tan ϕ'	—
p	Projection ratio	—
r_{sd}	Settlement ratio	—

g newtons = 9·806 N. In all the formulae, loadings expressed in kgf/m can therefore be converted to the equivalent N/m by multiplying by 9·806. An approximate value in N/m can be obtained by multiplying by 10·0.

To solve any specific problem, the following basic information is required:

1. Pipe diameter (nominal) in millimetres.
2. Trench width for design purposes in millimetres.
3. Maximum trench depth in millimetres.
4. Minimum trench depth in millimetres.
5. Density of fill material in kg/m^3.
6. Traffic loading—see p. 85.
7. Impact factor—see p. 85.
8. Safety margin to be used—see p. 85.

As both trench and pipe diameter have now been expressed in millimetres, the basic Marston formulae become:

Narrow-trench

$$W_c = C_d \gamma B_d^2 10^{-6} \text{ kgf/m}$$ **Formula No. 10.1**

Wide-trench

$$W_c' = C_c \gamma B_c^2 10^{-6} \text{ kgf/m}$$ **Formula No. 10.2**

In Formulae 10.1 and 10.2 the coefficients C_d and C_c are those originally proposed by Marston and Spangler. C_d is dependent on the ratio H/B_d and the product of K and μ'. Curves showing the values of this coefficient for various types of fill material are given in Fig. 10.3. The coefficient C_c is dependent on the ratio H/B_c and on the product of r_{sd} and p. Values of this coefficient for various values of $r_{sd}p$ are given in Fig. 10.4. These two figures are Crown Copyright and are reproduced by permission from the NBS Special Report No. 37. Where no specific information on soil conditions is available, the values of γ and $K\mu'$ given in Table 10.4 are generally satisfactory.

Under wide-trench conditions, for rigid pipe beddings (e.g. concrete) which do not extend over the full width of the trench, the settlement ratio (r_{sd}) and the projection ratio (p) may both be assumed as 1·0, giving a composite value of 1·0 for the expression $r_{sd}p$. When granular bedding is to be used over the full width of a 'wide' trench and up to the mid-diameter of a rigid pipe, however, the value of p may be taken as 0·5, so that $r_{sd}p$ then also becomes 0·5. When small-diameter pipes (up to 300-mm diameter) are laid on the natural trimmed trench bottom, the value of $r_{sd}p$ should be taken as 0·7. For all wide-trench conditions, $K\mu$ is generally taken as 0·190, as any lesser values would have very little effect on the value of coefficient C_c.

TABLE 10.4
SOIL DENSITIES AND COEFFICIENTS

Soil	*Density (γ), kg/m^3*	*Coefficient, $K\mu'$*
Sand and gravel	1 840	0·165
Saturated topsoil	1 920	0·150
Ordinary clay	2 000	0·130
Saturated clay	2 080	0·110

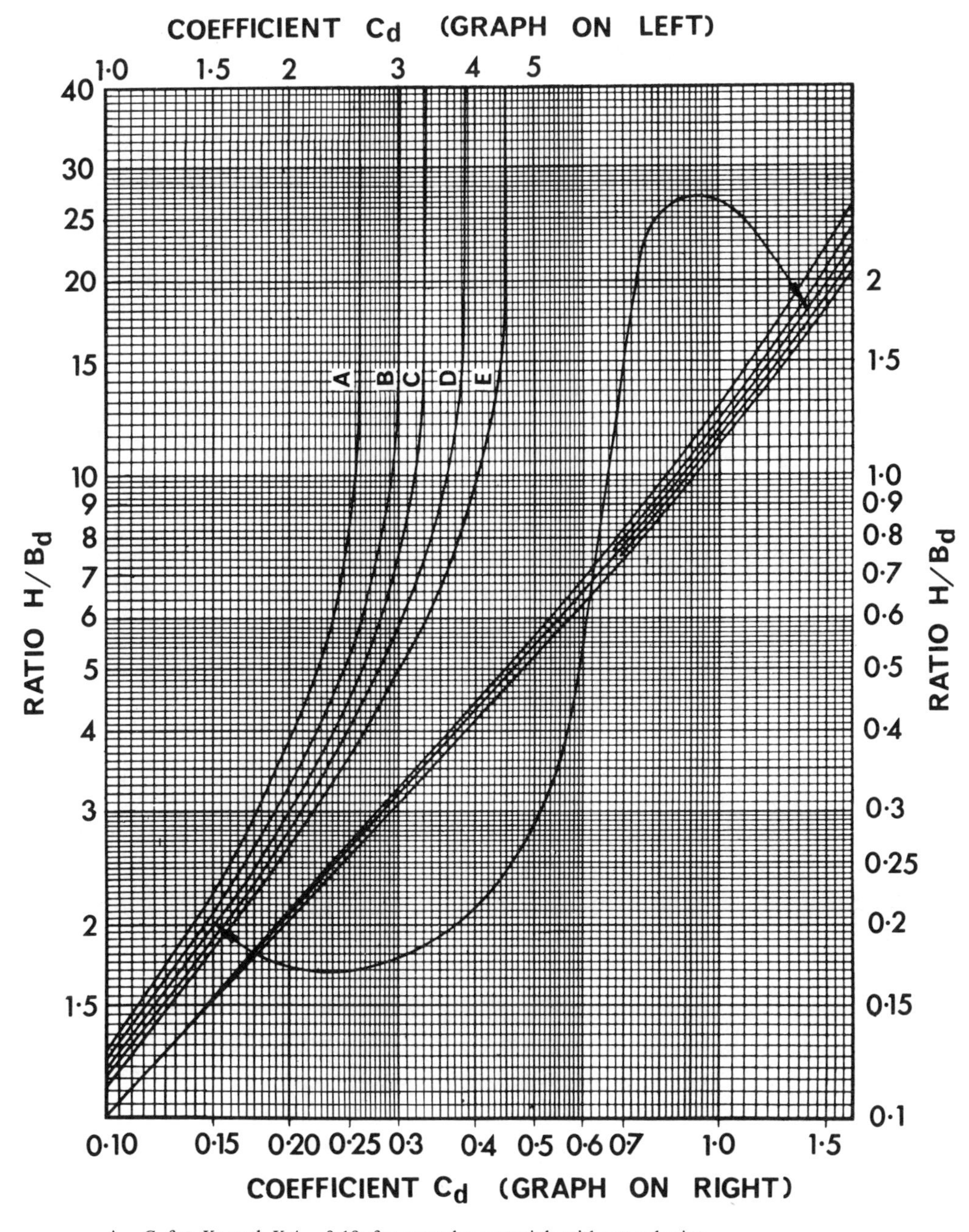

A—C_d for $K\mu$ and $K\mu'$ = 0·19, for granular materials without cohesion
B—C_d for $K\mu$ and $K\mu'$ = 0·165 max. for sand and gravel
C—C_d for $K\mu$ and $K\mu'$ = 0·150 max. for saturated top soil
D—C_d for $K\mu$ and $K\mu'$ = 0·130 ordinary max. for clay
E—C_d for $K\mu$ and $K\mu'$ = 0·110 max. for saturated clay

Fig. 10.3. *Narrow-trench fill load coefficients C_d. (Crown copyright. Reproduced from* Loading Charts for the Design of Buried Pipelines. *National Building Studies Special Report No.* 37.)

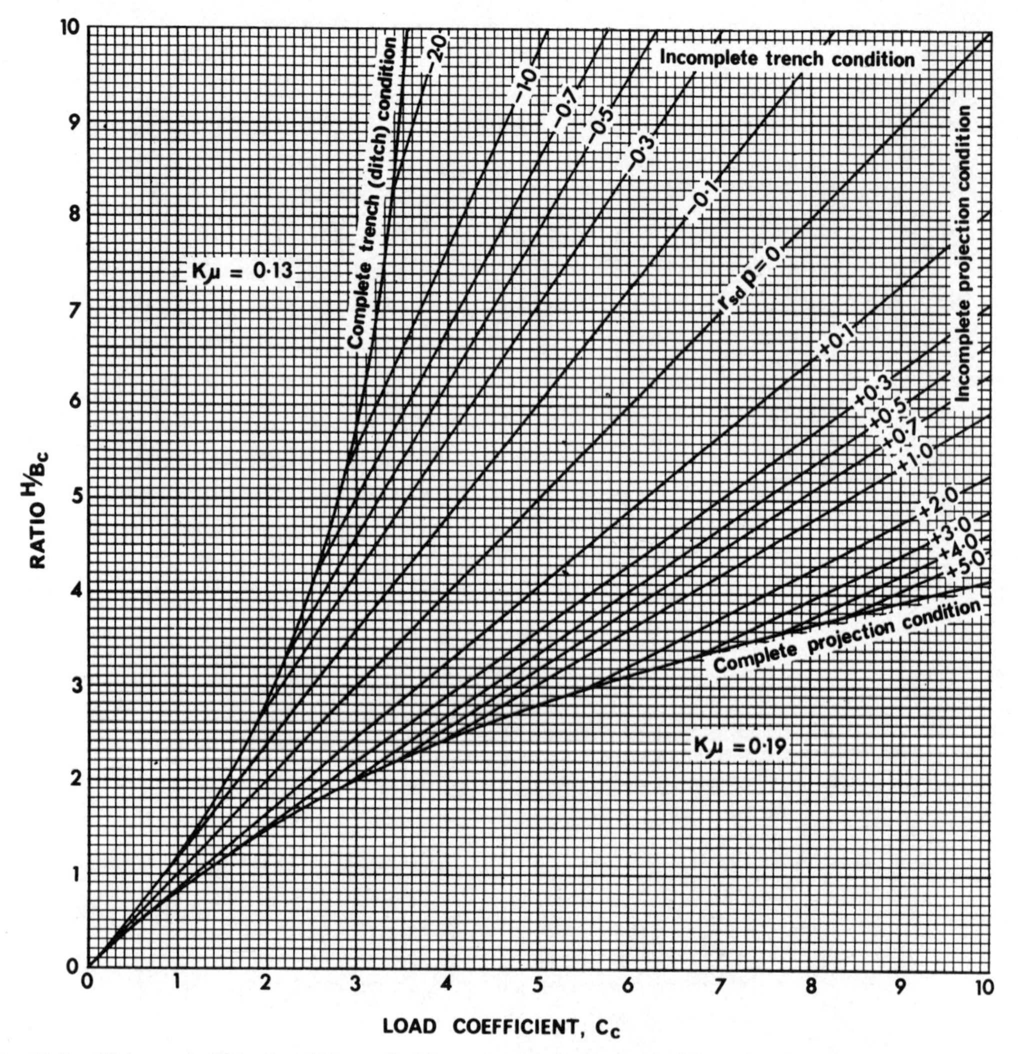

Fig. 10.4. *Wide-trench fill load coefficients C_c. (Crown copyright. Reproduced from* Loading Charts for the Design of Buried Pipelines. *National Building Studies Report No.* 37.)

For any specific pipeline, it is now possible to calculate comparable values of W_c and W'_c using either Formula 10.1 or Formula 10.2, along with suitable values of either C_d or C_c taken from Figs. 10.3 and 10.4. The *lower* of the two values (W_c or W'_c) will be used in later calculations of the total loading on the pipeline.

The curves in Figs. 10.5 to 10.10 give values of concentrated surcharge loads (W_{csu}) as recommended by the Working Party. These curves have been developed from Charts C13 to C16 of the NBS Special Report No. 37, and give values for the six possible conditions set out in Table 10.5. The curves in these six figures include allowances for the impact factors recommended by the Working Party.

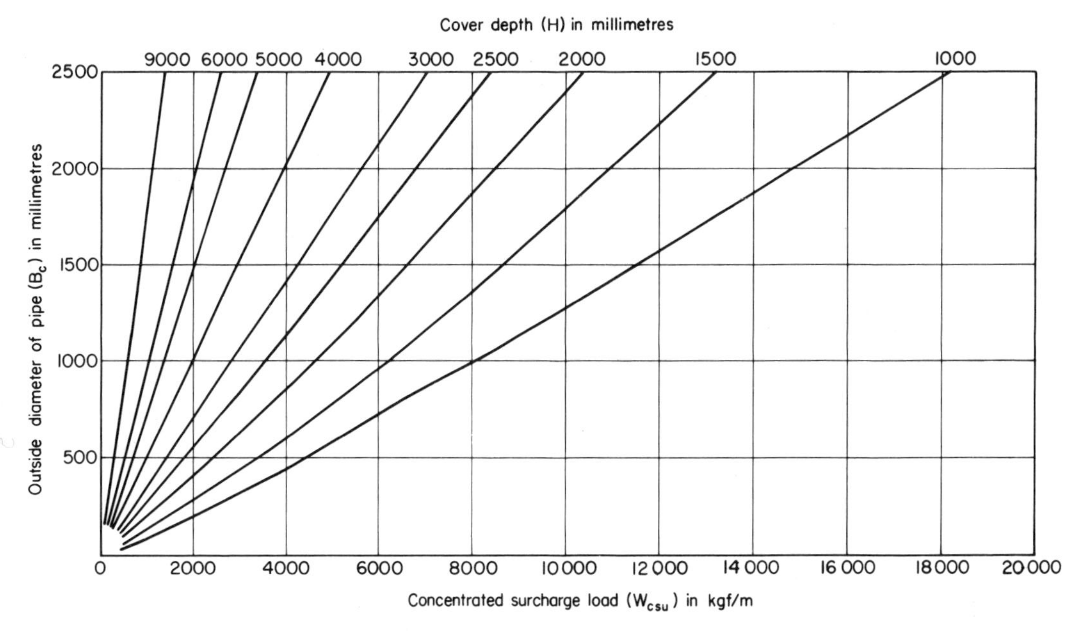

Fig. 10.5. *Concentrated surcharge loading—main traffic routes. Impact factor* 1·3, *Class 'B' bedding or concrete arch.*

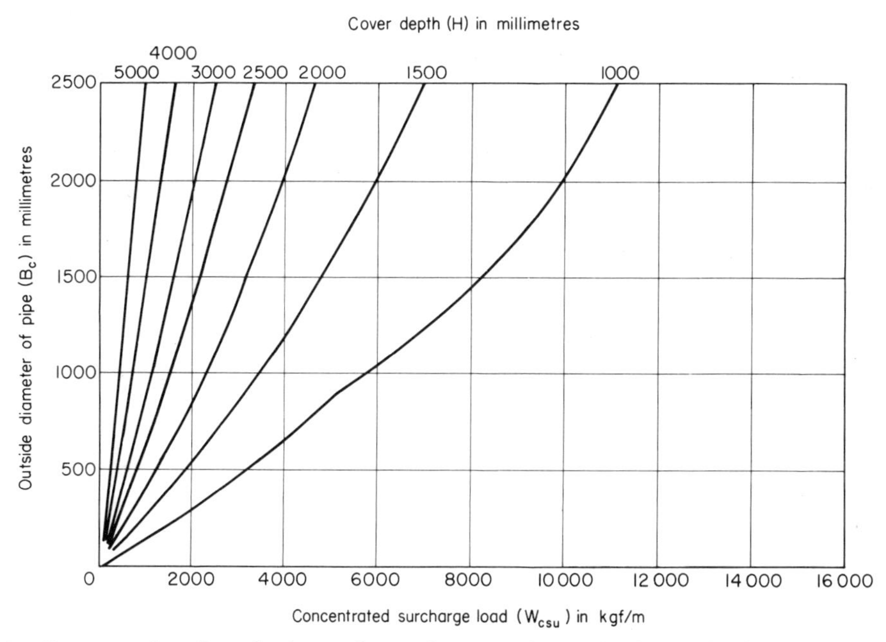

Fig. 10.6. *Concentrated surcharge loading—other roads. Impact factor* 1·5, *Class 'B' bedding or concrete arch.*

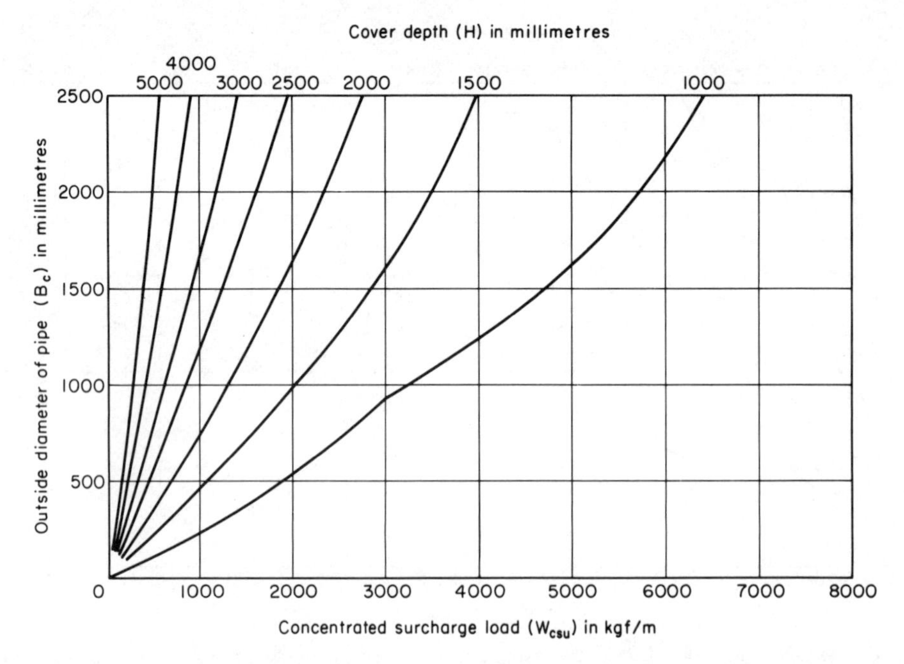

Fig. 10.7. *Concentrated surcharge loading—fields, gardens and lightly trafficked access roads. Impact factor* 2·0, *Class 'B' bedding or concrete arch.*

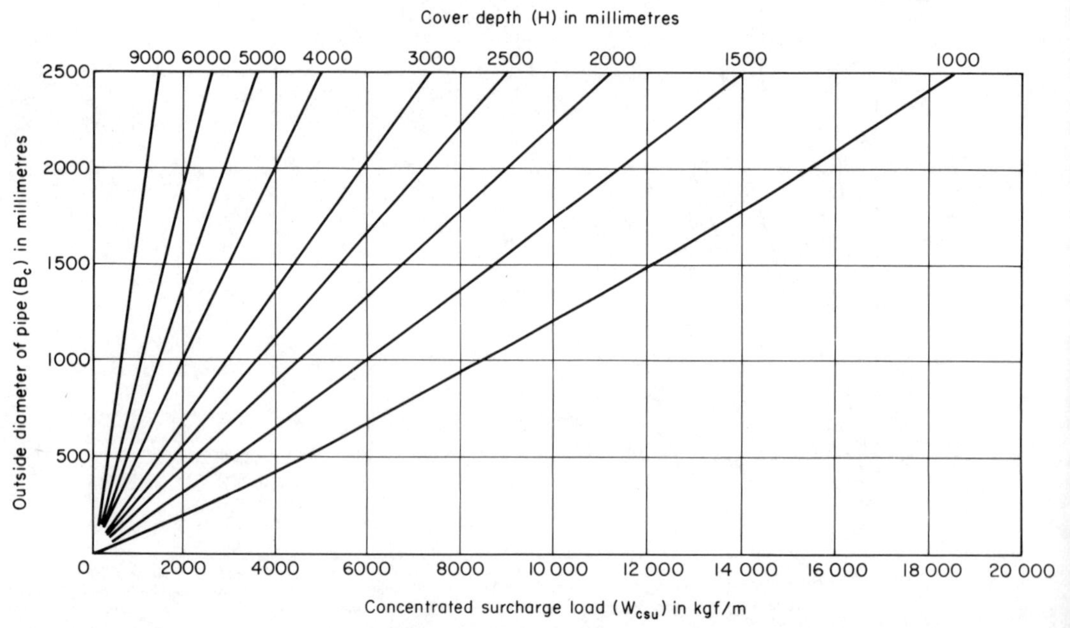

Fig. 10.8. *Concentrated surcharge loading—main traffic routes. Impact factor* 1·3, *Concrete cradle.*

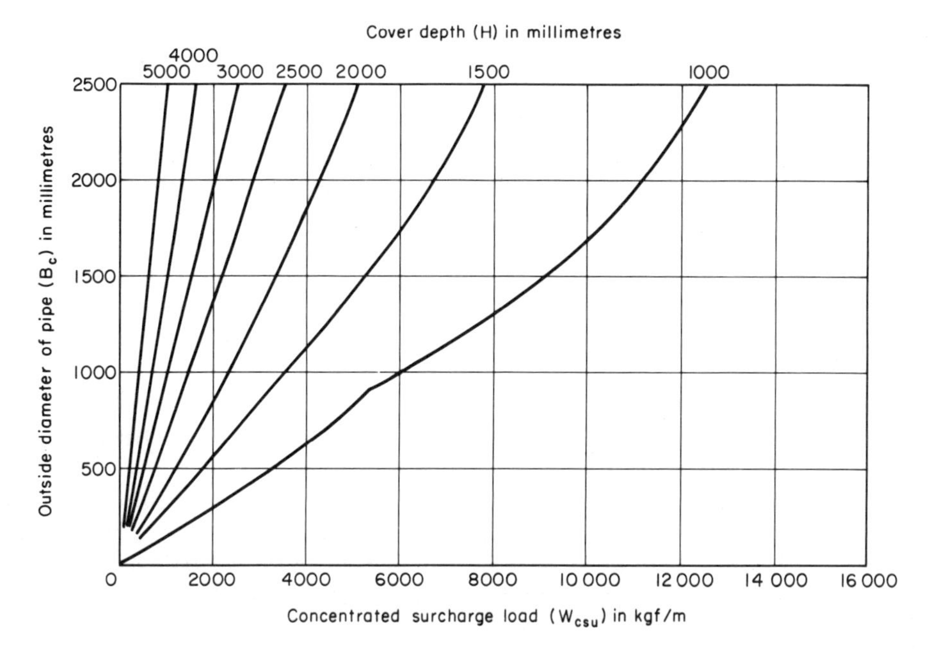

Fig. 10.9. *Concentrated surcharge loading—other roads. Impact factor* 1·5, *Concrete cradle.*

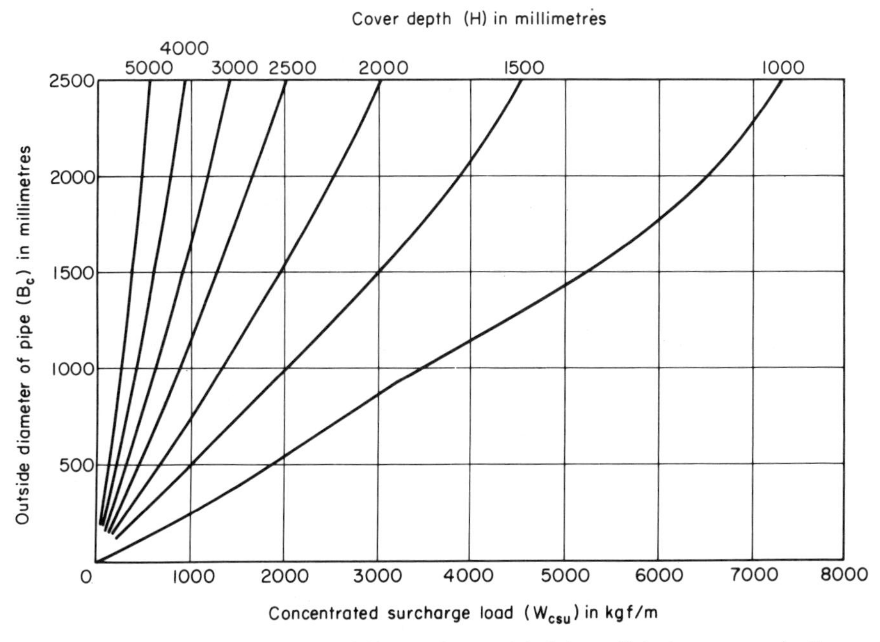

Fig. 10.10. *Concentrated surcharge loading—fields, gardens and lightly trafficked access roads. Impact factor* 2·0, *Concrete cradle.*

TABLE 10.5
CONCENTRATED SURCHARGE LOADS

Bedding type	*Traffic loading*	*Impact factor, (F_i)*	*Graph*
Class 'B' or Concrete Arch	Trunk roads	1·3	Fig. 10.5
	Other roads	1·5	Fig. 10.6
	Fields, etc.	2·0	Fig. 10.7
Concrete Cradle	Trunk roads	1·3	Fig. 10.8
	Other roads	1·5	Fig. 10.9
	Fields, etc.	2·0	Fig. 10.10

The first set of calculations should be carried out, using Figs. 10.5, 10.6 and 10.7. If it is then found that a Class 'B' bedding is not adequate, the calculations must be repeated, using Figs. 10.8 to 10.10, if it is proposed to use a concrete cradle, and not an arch.

For larger-diameter pipelines (over 300-mm diameter) it is usually advisable to add an additional allowance for the weight of the water in the pipes when calculating the total effective external load. Water weighs 1000 kg/m^3, and when the internal diameter is measured in millimetres, the equivalent external load on rigid pipes can be taken for design purposes as:

$$W_w = 0{\cdot}75\frac{\pi}{4}d^2 \times 10^{-3}\ \text{kgf/mm}$$

This can be simplified to:

$$W_w = 589d^2 \times 10^{-6}\ \text{kgf/m} \qquad \textbf{Formula 10.3}$$

The Working Party recommend that, except in special circumstances, no provision need normally be made for any distributed surcharge load over and above the allowance for traffic loading. The total effective distributed load for design purposes (W_e) would normally therefore be:

W_c or W_c'	(whichever is the lower)
plus W_{csu}	(the traffic loading)
plus W_w	(if applicable)

The strength of the pipeline itself (W_f) is calculated from the value of W_T (the safe crushing-test strength of the pipes), multiplied by a suitable safety margin F_s when the pipes are unreinforced (recommended as 0·80 by the Working Party), and also multiplied by the bedding factor F_m (see Table 10.1).

The total effective load on the pipeline (W_e) must, of course, be less than the strength of the pipeline (W_f). If the first calculation of W_f is too low, either:

i. stronger pipes must be used (increase W_T),

or

ii. a stronger bedding must be specified (increase F_m).

If the proposed type of bedding is changed from granular (type 'B') to one of the concrete 'cradle' protections, the charts in Figs. 10.8 to 10.10 must be used to calculate a new value for W_{csu}: this may then entail a slight adjustment in the value of W_e.

The advantages on site from using an arch instead of a cradle for concrete protection were discussed earlier in this chapter. When it is proposed to use an arch for Class 'A' or 'A_{rc}' bedding, two factors must be taken into account:

i. the value of 'H' must be taken as the height of cover over the concrete arch, and *not* over the pipe itself, and
ii. the value of B_c in wide trench calculations must be taken as the actual width of the top of the concrete, and *not* the outside diameter of the pipes.

It will be apparent that item (i) above will normally have only a very minor effect on the calculations, but that by increasing the value of B_c the use of concrete arch protection under wide-trench conditions would increase the value of W_e. This will only be detrimental when this would then make the value of W_e greater than W_f.

TUNNELS, HEADINGS AND THRUSTBORES

For pipes in tunnels and headings, the Marston trench formulae can be used. Although modification to include a factor for cohesion of the soil is not proposed by Clarke [67], it is included in the American Water Pollution Control Federation Manual No. 9 [84]. If this factor is taken into account, the formula becomes:

$$W_t = C_t B_t (\gamma B_t - 2c) \cdot 10^{-6} \text{ kgf/m} \qquad \textbf{Formula 10.4}$$

where
W_t is the fill load on the pipeline or on the tunnel support
B_t is the overall width of the tunnel or heading
c is the coefficient of cohesion in kg/m² (see Table 10.6)

TABLE 10.6
COEFFICIENT OF COHESION—IN HEADINGS

Soil	*Coefficient in kg/m²*
Very soft clay	200
Medium clay	1 200
Hard clay	5 000
Loose, dry sand	0
Silty sand	500
Dense sand	1 500

In the above formula C_t is the load coefficient, and is a function of H/B_t and the coefficient of internal friction of the soil. It is the same as coefficient C_d used in narrow-trench calculations (see Fig. 10.3), except that H is measured either to the top of the pipeline in heading or to the top of the tunnel lining, and that B_t is used in lieu of B_d.

Values of the coefficient of cohesion would normally be ascertained from laboratory tests on the soil. If these are not available, the figures set out in Table 10.6 can be considered as 'safe' values.

When pipelines are constructed by thrustboring, the fill loading can be calculated from Formula 10.1, using the external diameter of the pipeline for B_d. NBS Special Report No. 37 states that the value of F_m for thrustbore conditions requires experimental investigation, and that it may be assumed tentatively as 1·5.

FLEXIBLE PIPELINES

While much of the design practice for rigid pipelines will apply also to flexible pipes, it must be borne in mind that the strength of a line of flexible pipes depends to a great extent on the effectiveness of the sidefill. The term 'flexible pipes' embraces pipes of PVC and pitch-fibre, together with those of ductile iron and steel (including corrugated steel).

The Marston formulae can be adjusted for use with flexible pipes as follows:

i. Narrow trench

$$W_c = C_d \gamma B_c B_d 10^{-6} \text{ kgf/m} \qquad \textbf{Formula 10.5}$$

ii. Wide trench

$$W_c' = C_c \gamma B_c^2 10^{-6} \text{ kgf/m} \qquad \textbf{Formula 10.6}$$

The bedding and sidefill to flexible pipes must be of approved granular material, and should extend at least up to 100 mm above the crown of the pipes to provide the necessary side support. Adequate compaction of the backfill is essential, and the deflection of the outside diameter of the pipes should be limited to a maximum of 5%.

Knowledge of this subject is limited and methods of calculations of the deflection are to a great extent empirical. For pipelines buried in properly controlled compacted backfill, Clarke [67] has proposed a simplification of the formula put forward earlier by Spangler, so that for thin-walled pipes:

$$\Delta_x = \frac{2{\cdot}7 p_0}{k} \qquad \textbf{Formula 10.7}$$

where/

Δ_x is the deflection in millimetres

p_0 is the unit pressure at top of pipe in kgf/m^2

k is a coefficient of soil reaction, which for pipes up to 1500-mm diameter, should be not less than 550 kgf/m^2/mm

SHALLOW PIPELINES

Where pipes are laid with less than about 1·20 m of cover in roads, special protection will be needed against traffic loading and vibration. This particularly applies to gully connections. The author prefers to surround the gully pot and its immediate connecting pipe with concrete, and then to provide a flexible joint on the pipeline immediately outside the concrete. Very shallow pipes can be protected by a reinforced concrete slab, spanning the trench and supported on the undisturbed soil at each side of the trench. Where the pipeline is so shallow that this slab might rest on the pipes themselves, pipes of steel or ductile iron should be used.

Pipes under fields and gardens should preferably have at least 0·75 m of cover to avoid damage during ploughing, etc. If this is not possible, the pipes should be covered with suitable slabs or tiles.

TYPICAL CALCULATIONS

Example 1. A 150-mm diameter vitrified clay pipeline is to be laid in wide-trench conditions in sand and gravel with a depth to invert of 2·00 m. An unclassified road will be constructed over the pipeline later.

Assume a granular bedding to the full width of the excavation, then,

$$r_{sd}p = 0{\cdot}5$$

the external diameter

$$B_c = 190\,\text{mm}$$

soil density

$$\gamma = 1840\,\text{kg/m}^3$$

As

$$H_i = 2000\,\text{mm}$$
$$H = 2000 - 150 \text{ (approx.)}$$
$$= 1850\,\text{mm}$$

and,

$$H/B_c = 9{\cdot}75$$

(*a*) *To calculate the fill loading:*

Taking $K\mu$ as 0·19,

From Fig. 10.4

$$C_c = 14{\cdot}55$$

From Formula 10.2

$$W_c' = 14{\cdot}55 \times 1840 \times 190^2 \times 10^{-6}$$
$$= \mathbf{966\,kgf/m}$$

(*b*) *To calculate the traffic loading:*

From Fig. 10.6

$$W_{csu} = 250\,\text{kgf/m}$$

(*c*) *The total effective design loading is then:*

$$W_e = W_c' + W_{csu} = 964 + 250$$
$$= \mathbf{1214\,kgf/m}$$

(*d*) *To calculate the strength of the pipeline:*

The safe crushing test strength of a 150-mm 'extra strength' vitrified clay pipe to BS 65 and 540

$$= W_T = 2\,200\ \text{kgf/m}$$

Taking a safety margin of 0·80, and a bedding factor of 1·9 (see Table 10.1):

$$\begin{aligned} W_f &= W_T \times F_s \times F_m \\ &= 2200 \times 0{\cdot}8 \times 1{\cdot}9 \\ &= \mathbf{3344\ kgf/m} \end{aligned}$$

This is greater than the effective design loading calculated above, and the structural design is therefore satisfactory.

Example 2. A 900-mm diameter concrete pipeline is to be laid across a field in a trench 1500 mm wide, at a depth of 4·00 m to the invert. The subsoil is expected to be saturated clay.

(*a*) *To calculate the fill loading* (*narrow trench*)*:*

Width of trench,

$$B_d = 1500\ \text{mm}$$

soil density

$$\gamma = 2800\ \text{kg/m}^3$$
$$H_i = 4000\ \text{mm}$$

then

$$\begin{aligned} H &= 4000 - 900\ \text{(approx)} \\ &= 3100\ \text{mm} \end{aligned}$$

and

$$H/B_d = 2{\cdot}06$$

From Fig. 10.3 (curve *E*)

$$C_d = 1{\cdot}65$$

From Formula 10.1

$$\begin{aligned} W_c &= 1{\cdot}65 \times 2080 \times 1500^2 \times 10^{-6} \\ &= \mathbf{7722\ kgf/m} \end{aligned}$$

(*b*) *To calculate the fill loading* (*wide trench*)*:*

let external diameter

$$B_c = 1050\ \text{mm}$$

soil density

$$\gamma = 2080\ \text{kg/m}^3$$

as before

$$H = 3100\ \text{mm}$$

and

$$H/B_c = 2{\cdot}95$$

assuming $r_{sd}p = 0{\cdot}5$, and $K\mu = 0{\cdot}19$

From Fig. 10.4

$$C_c = 4{\cdot}35$$

From Formula 10.2

$$W'_c = 4{\cdot}35 \times 2080 \times 1050^2 \times 10^{-6}$$
$$= \mathbf{9975\ kgf/m}$$

(*c*) *To calculate the traffic loading:*

From Fig. 10.7

$$W_{csu} = 650\,\text{kgf/m}$$

(*d*) *The total effective loading on the pipeline is then:*

$$W_e = W_c \text{ (as this is lower than } W'_c) + W_{csu}$$
$$= 7730 + 650$$
$$= \mathbf{8380\ kgf/m}$$

(*e*) *The strength of the pipeline:*

Assuming class 'M' unreinforced pipes on a granular bed,

$$W_T = 6850\,\text{kgf/m}$$

then

$$W_f = 6850 \times 0{\cdot}8 \times 1{\cdot}9$$
$$= \mathbf{10\,400\ kgf/m}$$

This is satisfactory.

SIMPLIFIED DESIGN CHARTS

Design charts 1 to 9 (see pages 182–190) have been prepared to obviate the need for calculations for small-diameter pipelines. When decisions have been made on the type of subsoil to be expected, the loading classification, and the safety margin to be employed, the relevant chart can be selected according to pipe diameter and the design trench width. The charts are suitable only for use with vitrified clay pipes to BS 65 and 540 'Extra Strength' quality, and should not be used for pipes of other materials.

The charts originally formed part of a paper given to the Institution of Public Health Engineers in 1967 [42], and were revised in June 1968 [85], after the publication of the Second Report of the Ministry Working Party. They include all the design criteria recommended by the Working Party.

From the relevant chart of pipe diameter and trench width, the total loading on the pipeline (in either kgf/m or in N/m) can be read off directly if required, according to the depth to invert and the type of subsoil. Alternatively, the required bedding factor can be read off from the right-hand side of the charts, depending on the safety margin to be adopted.

Example. A 225-mm pipeline of 'extra strength' vitrified clay pipes is being laid on a class 'B' bed in a trench of sandy gravel at a depth of 4·5 m to invert. A slip in the trench sides has resulted in a maximum width of about 2·00 m at the crown of the pipes. Trunk-road loading is expected. Is the pipeline satisfactory, using a safety margin of 0·80?

(*a*) Chart No. 6 can be used for *any* trench width, as the transition depth is beyond 6·0 m.
(*b*) On Chart No. 6 locate the depth to invert of 4·5 m.
(*c*) Draw a vertical line to cut the 'sand and gravel' curve for trunk-road loading.
(*d*) The total design loading is given on the left-hand scale as 3875 kgf/m (38 014 N/m).
(*e*) Draw a horizontal line to cut the 'safety margin' curve for 0·80.
(*f*) Draw a vertical line down to the base. This cuts the bedding factor scale at about 1·80.

This pipeline will therefore be satisfactory on a granular bedding, despite the slip in the trench during construction.

Design charts for the structural design of pipelines have been published by *Municipal Engineering* in their series of data sheets [83]. These cover the use of vitrified clay, concrete and asbestos-cement pipes up to 450-mm diameter for a soil density of 2000 kg/m^3 only. The Clay Pipe Development Association has issued sets of design tables [86] for use with vitrified clay pipes up to 450-mm diameter, for various trench widths, and for depths to 6·00 m.

11 Manholes and Other Ancillary Works

MANHOLES are provided on sewers and drains as a means of access for inspection and testing, and for the clearance of any obstructions. On house drainage these chambers are generally referred to as inspection chambers. Except for very shallow drains and sewers of less than about 1 m depth to invert, all manholes should be of adequate dimensions for entry and for the operation of drain rods. A manhole should always be fitted with stepirons (or a ladder if very deep), for ease of both entry and exit.

It is usual to construct a manhole at every change of direction or change of gradient on all sewers which are not large enough for a man to enter, i.e. on all sewers of up to about 900-mm diameter. To allow for normal methods of rodding, manholes are usually provided at not more than about 100 to 110 m intervals, while some engineers prefer a closer spacing. It may sometimes be found that any slight saving in cost by the introduction of a change in gradient may in fact be more than offset by the cost of the extra manhole.

On larger sewers (900-mm diameter and over) manholes should be provided at or near every material change of sewer size or gradient, and at major junctions. They should in any case not normally be more than about 200 m apart, so that men working in a sewer can easily reach a manhole in an emergency. In certain circumstances BSCP 2005 approves of manholes at intervals of up to 360 m on very large sewers (1800 mm and over), where these are laid in heading or tunnel. Otherwise, the Code recommends a spacing of 150 m per 1000 mm of diameter.

The requirements of the Water Authorities are that manholes are provided at all changes of direction and gradient, and at distances apart not exceeding about 100 m, but that on sewers in which men can work 'this distance may be increased'. The Ministry's Working Party considered that, in view of the substantial cost of manholes, the whole question of manhole spacing and its relationship with sewer lines and sewer maintenance was one 'worthy of investigation'. In the meantime they have recommended that no change be made, 'except that no objection be raised to the use of longitudinal curves of reasonable radius on sewers in which men can work'. It was pointed out that this has in fact been accepted practice for some time.

Building Regulations 1976, require an inspection chamber on a drain within 12·5 m of any junction (unless there is a chamber at the junction itself), and also at the highest point on a private sewer unless there is a rodding eye. No part of a drain or private sewer may be more than 45 m from an inspection chamber on that drain or private sewer.

MANHOLE DIMENSIONS

On pipelines up to 300-mm diameter, rectangular brick or *in situ* concrete manholes should be at least 1350 mm long by 788 mm wide for sewer depths to about 3 m to invert (these figures have been chosen to fit standard brick sizes). For deeper sewers to about 8 m, it is normal to have an access shaft of at least 788 by 675 mm leading to a lower chamber. This lower chamber should preferably

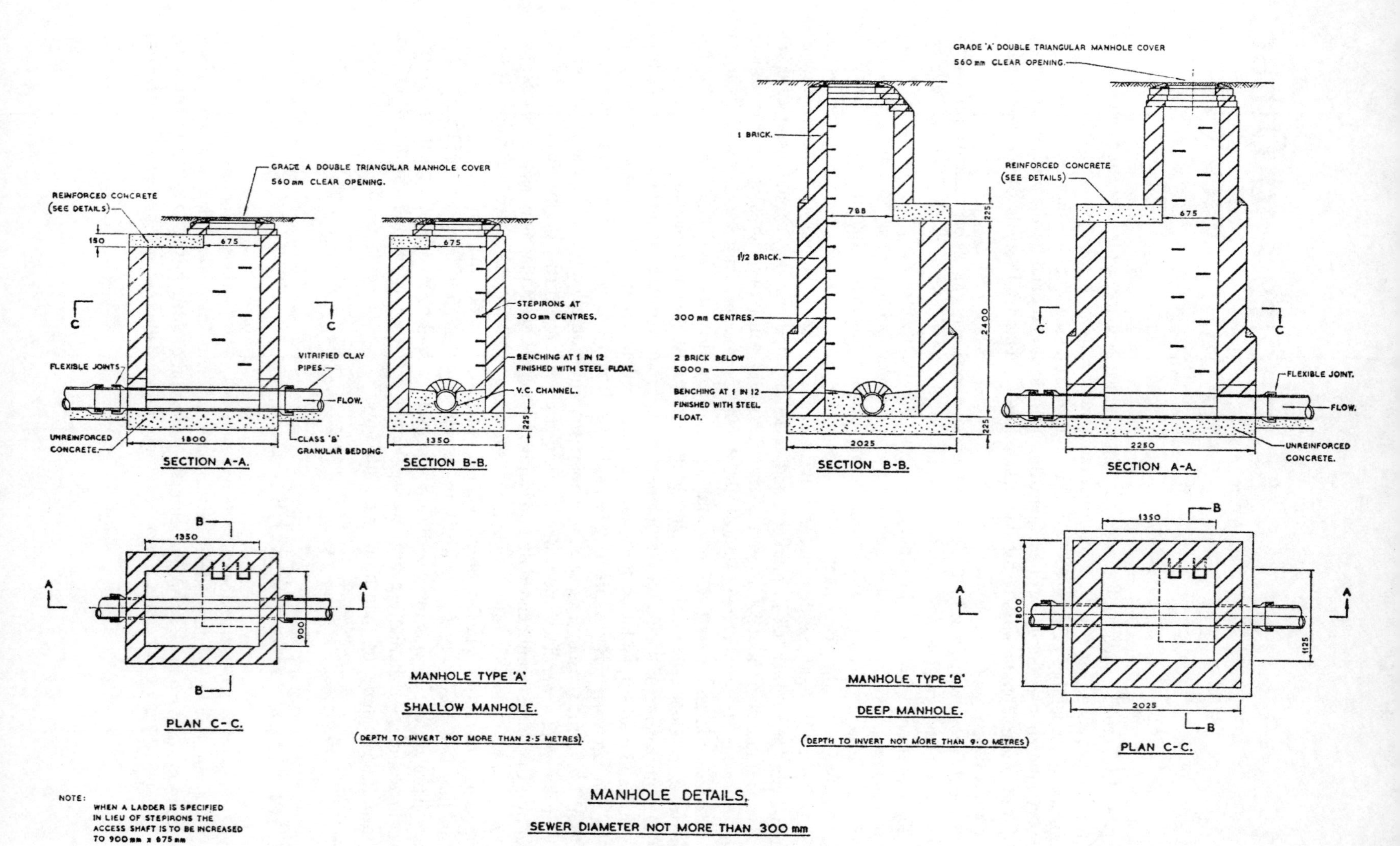

Fig. 11.1. *Typical manhole detail—in brickwork.*

TABLE 11.1
MAXIMUM DEPTH OF BRICK MANHOLES (METRES)

Wall length (mm)	*Wall thickness (mm)*			
	225	*337*	*450*	*562*
675	9·0			
900	5·5	11·5		
1 125	3·5	7·5		
1 350	2·5	5·0	10·0	
1 575		4·0	7·5	11·5
1 800		3·0	5·0	8·0
2 025		2·5	4·5	6·5
2 250			3·5	5·0
2 475			3·0	4·5
2 700			2·5	3·5

have a minimum clear height of 2·0 m above the benching (see CP 2005), and should generally be at least 1350 mm long by 1125 mm wide. Some Water Authorities now require the shaft to be not less than 900 mm by 600 mm. A typical manhole in 225-mm thick brickwork is shown in Fig. 11.1.

While very shallow brick manholes and inspection chambers can be constructed in half-brick thickness (112 mm), it is usual to make manhole walls at least one-brick thickness (225 mm). It is necessary to use thicker walls as the length and depth of the walls increase. Table 11.1 gives approximate maximum depths of brick manholes and other structures for various wall spans.

Manholes on larger diameter sewers must be wide enough to accommodate the wider channels plus the benching. Suggested widths for both shallow and deep manholes are set out in Table 11.2.

Manholes are frequently constructed of precast concrete tubes. These are specially manufactured for manhole construction, complete with precast inverts, taper sections, shaft sections and cover slabs. Chamber diameters will vary with sewer diameter and are available from 900 to 1 800 mm, in 150-mm increments, while shafts are standardized at 675-mm diameter. Some Water Authorities now require the shaft to be not less than 900 mm diameter. Shafts, tapers and main segments are normally supplied complete with stepirons fixed in position at 300-mm centres.

TABLE 11.2
MINIMUM WIDTHS OF MANHOLES ON LARGER SEWERS

Sewer diameter (mm)	*Inside width of manhole (mm)*
375	1 125
450	1 125
525	1 238
600	1 350
675	1 463
750	1 575

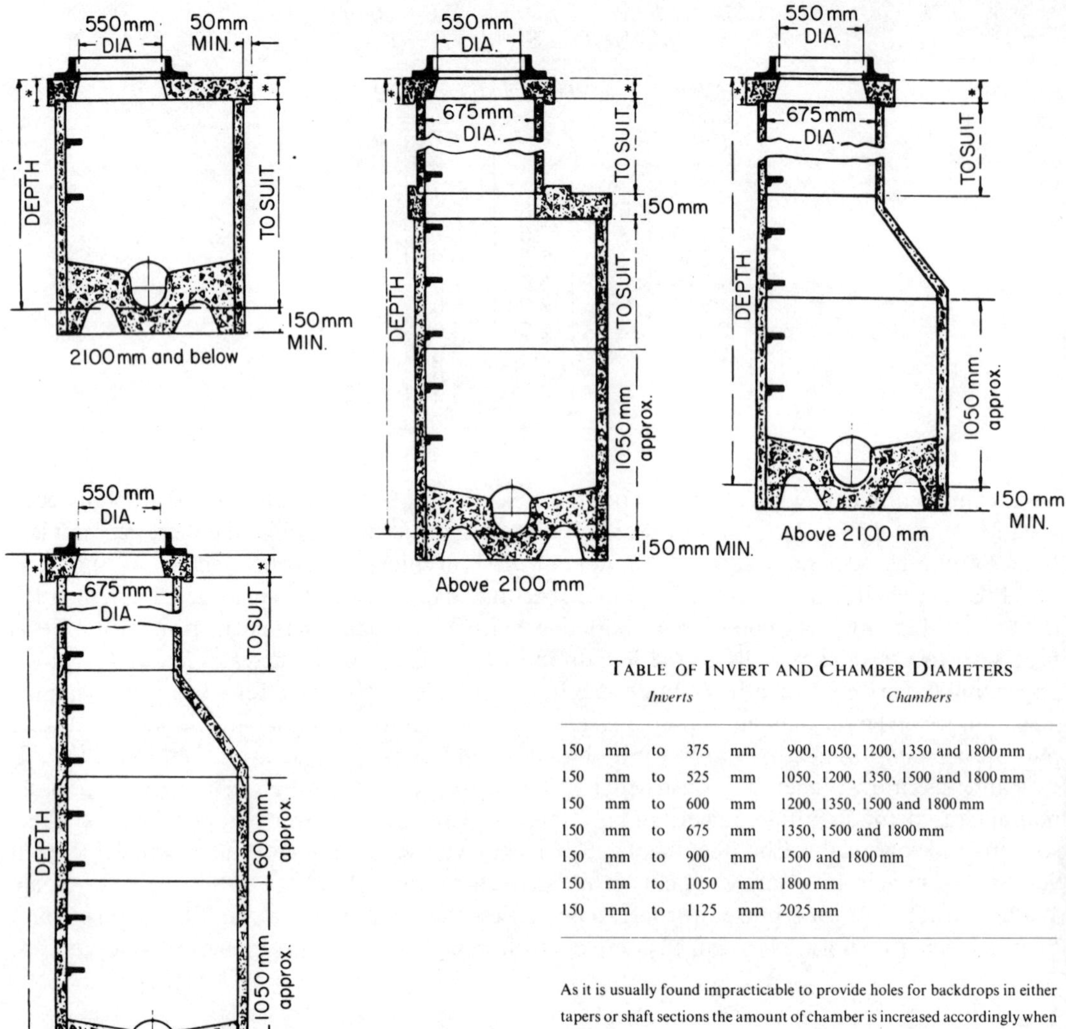

TABLE OF INVERT AND CHAMBER DIAMETERS

Inverts	*Chambers*
150 mm to 375 mm	900, 1050, 1200, 1350 and 1800 mm
150 mm to 525 mm	1050, 1200, 1350, 1500 and 1800 mm
150 mm to 600 mm	1200, 1350, 1500 and 1800 mm
150 mm to 675 mm	1350, 1500 and 1800 mm
150 mm to 900 mm	1500 and 1800 mm
150 mm to 1050 mm	1800 mm
150 mm to 1125 mm	2025 mm

As it is usually found impracticable to provide holes for backdrops in either tapers or shaft sections the amount of chamber is increased accordingly when necessary.

All shaft and chamber sections, and manhole tapers, have galvanised malleable iron stepirons, which comply with BS 1247, built in at the works.

Lifting holes are provided in sections above 686 mm diameter.

Ring bolts may be loaned if required.

For chamber sizes up to 1200 mm, tapers are 600 mm deep, for chamber sizes over 1200 mm, tapers are 900 mm.

Fig. 11.2. *Typical manhole detail—in precast concrete (by courtesy of Stanton and Staveley Ltd).*

As these manhole segments have ogee joints, they should normally be surrounded with 150-mm thickness of good quality concrete to ensure that they will be watertight. Figure 11.2 shows a typical arrangement for a manhole constructed of precast concrete segments. In waterlogged ground, the weights of the manholes should be checked to ensure that there will be no tendency to float, during construction or after completion.

BACKDROP MANHOLES

To avoid the cost of deep excavation for sewers with steep gradients, it is often more economical to lay a drain or sewer at a gradient sufficient for the hydraulic requirements, and then to connect this to a lower sewer by means of a backdrop manhole (see Fig. 11.3). The incoming drain or sewer is led into a vertical pipe (normally constructed immediately outside the manhole), which in turn terminates at its lower end in a 90° bend at or just above the invert level of the lower sewer. Access is provided through the manhole wall to the upper level drain for inspection or rodding. The tee-piece, vertical pipe and bend should be surrounded with at least 150-mm thickness of concrete.

BSCP 301 recommends that a backdrop manhole should be used if the difference in levels of the two drains is more than 600 mm. If the difference is less than this, it can be taken up by using a ramp formed in the benching. CP 2005 recommends that for differences in depth up to 1800 mm, a 45° ramp should be formed in the last part of the upper sewer, or in the benching. If the ramp is formed in the sewer, it should take the form of a 45° junction pipe with the main upper pipeline extended into the manhole wall as before for inspection and rodding.

CHANNELS AND BENCHINGS

Where a pipe passes through a manhole wall or connects with a backdrop, it should preferably be provided with a flexible joint as close to the outer face of the wall or concrete as possible. For smaller sewers (up to 300-mm diameter) this can conveniently be arranged by building in a special short length of pipe, about 300 mm in length, with a flexibly jointed socket just outside the wall. This is illustrated in Fig. 11.1. Where precast concrete rings are used, the pipes may be built into a lower *in situ* section, as shown in Fig. 11.2, or they can be connected into specially made precast base sections which are supplied with holes cast ready for pipe entries.

In brick manholes it is usual to build relieving arches over the pipes where they pass through the walls. These may be half-brick or one-brick for shallow manholes and for sewers up to 300-mm diameter, but they are usually of two-brick thickness in deep manholes when the sewers are over 300-mm diameter.

Channels in manholes should preferably be formed in half-round channel pipes of the same material as the sewer pipes. Dimensions of vitrified-clay channels and side branches are set out in BS 65 and 540, and BS 539. Channels may also be formed of *in situ* concrete with a granolithic finish.

Benchings should be constructed at each side of the channel, rising vertically from the mid-diameter to at least the height of the soffit of the sewer, and then sloping up to the sides of the manhole. CP 2005 recommends a fall on the surface of the benchings of about 1 in 36, but many

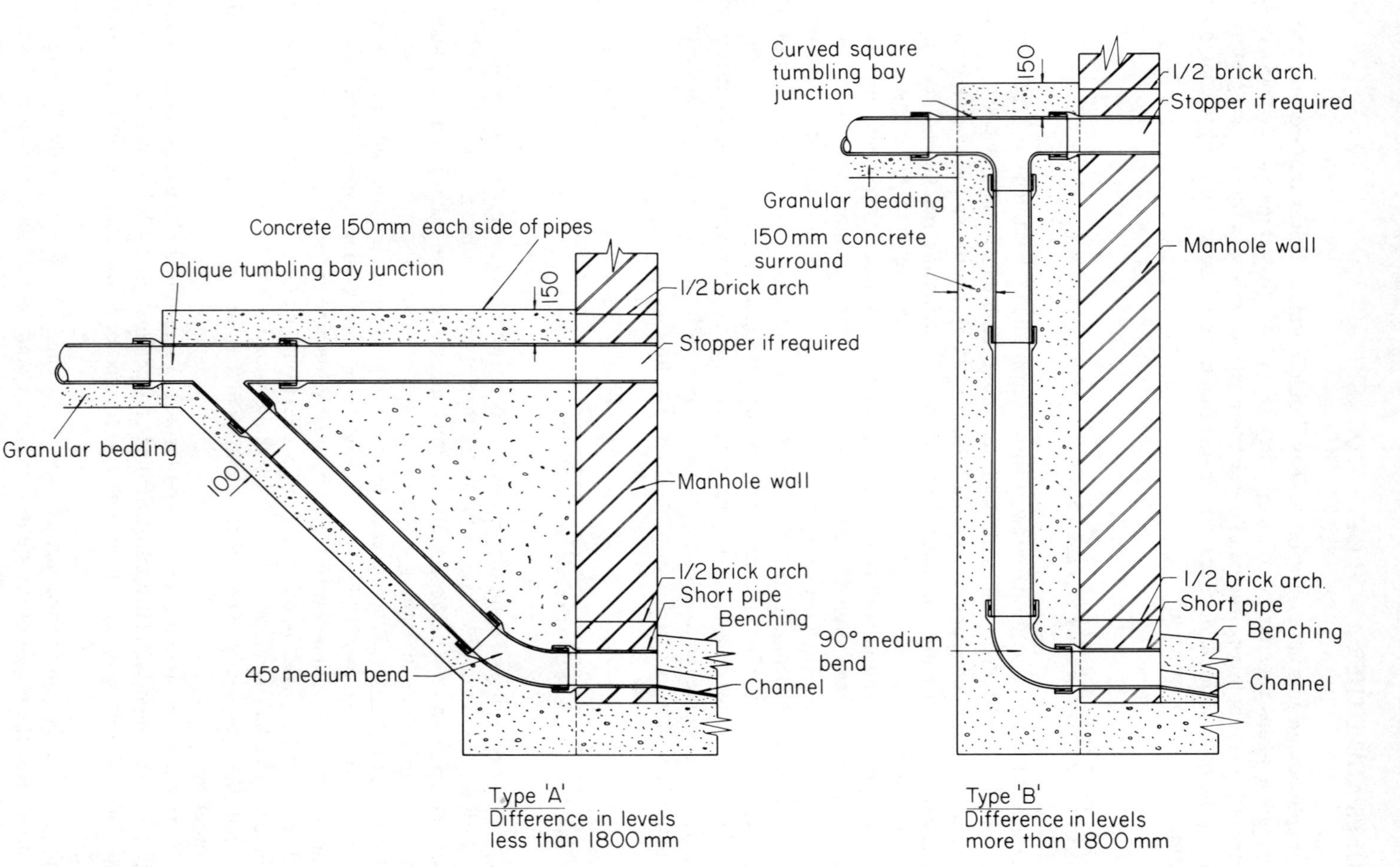

Fig. 11.3. Typical backdrop manhole. Note: rigid joints may be used in lieu of flexible joints where these are surrounded with concrete.

engineers prefer a steeper slope—1 in 12 or even 1 in 6, joining the vertical edge with a nosing of 20- to 25-mm radius. The sloping and vertical faces are preferably finished off with a steel float to form a hard, smooth surface, using 1:1 cement/sand mortar. CP 2005 recommends that benchings should be at least 230-mm wide on each side of the channel. In manholes on sewers of 375-mm diameter and over, one benching should be at least 350-mm wide to provide enough room for a man to stand. When the sewer diameter is 900 mm or over, safety chains should be provided to prevent a man from being swept down the sewer in times of storm.

ACCESS TO MANHOLES AND OTHER CHAMBERS

Except at very deep manholes, access is normally provided by stepirons. These should be to BS 1247 and are set at 300-mm vertical intervals, staggered at 225- or 300-mm horizontal centres. Chamber rings to BS 556 have stepirons set at 300-mm horizontal centres with the top stepiron not more than 350 mm below the underside of the manhole slab. Allowing for normal construction, with two courses of brick packing between the slab and the manhole cover, this means that the first stepiron will normally be not more than 800 mm below ground level.

For deep manholes (over about 4·5 m depth) ladders of heavy pattern galvanized steel are preferable to stepirons. They should be fixed to the sides of the shaft and manhole (and not to the benching), using gunmetal bolts. When a ladder is used, the shaft should be slightly wider than for stepirons; it is recommended that this should not be less than 900 by 675 mm. Further information on ladder details can be found in CP 2005.

Manhole covers should be to BS 497, and should always have a clear opening of at least 550 mm (some authorities require clear openings of 600 mm and these are now included in BS 497). Covers may be rectangular, circular, triangular, or double triangular, according to preference, and they should be of a strength to suit the traffic loadings to be expected. Many engineers now prefer to use either single or double triangular covers, as these are easy to open, and they do not rock once the seating begins to wear. Various patent-type covers are available, including those with locking facilities and those recessed for filling with concrete or other floor finishes. Building Regulations, 1976, require that covers of chambers within a building should be of the non-ventilating type with airtight seal, secured to the frame by removable bolts of corrosion-resistant material. Wherever possible, the frames of manhole covers should be bedded on one or two courses of brickwork (engineering bricks), to facilitate the setting of the cover to match the camber of the road surface, and to allow for any future adjustment in its level (up or down) without having to damage the cover slab of the manhole itself.

Manholes in agricultural land should not normally have their covers set at ground level. They should either be brought up to about 600 mm above natural ground level and mounded round with earth, or the chamber (or shaft) should be stopped off below ground level and covered with a suitable slab. The top surface of a buried cover slab should be from 450 to 600 mm below the surface of the ground.

FLUSHING

Although flushing tanks were frequently incorporated into sewerage systems in the past, they are now rarely used. Flushing, when necessary, can be carried out from a water cart or hydrant through a normal manhole. Sewers designed and laid to proper gradients should not require flushing. This will normally only be necessary in the upper lengths of sewers which have been designed to cater for future development, and which in the initial stages may be either too large or too flat. Flushing in these circumstances is best accomplished by fitting a disc-type penstock at the outlet of the top manhole, so that the manhole itself can be filled with water, which can be released quickly.

Automatic flushing tanks may sometimes be necessary where the sewerage system includes old sewers laid at flat gradients. A flushing tank for these conditions is usually a specially constructed chamber incorporating an automatic flushing siphon continuously fed from a source of water. A spring or other natural water source can sometimes be used. Where the public water mains are to be used, the Water Authority normally requires that the supply shall be through a bibtap discharging so that the tap is never in contact with the water level in the chamber. The tap should be in a separate compartment connected through a trap to the chamber containing the siphon. The capacity of the flushing tank should, if possible, be about 10 % of that of the sewer to be flushed.

VENTILATION OF SEWERS

Ventilation of sewers and drains is necessary to ensure a free circulation of the air through the system, and to prevent the build-up of foul air or poisonous gases. In the past, when interceptors were fitted to all house drains, the sewerage system was ventilated separately through special vent columns, connected to the upper manhole of each line of sewer.

Where existing properties have drains fitted with interceptors, vent columns may still be necessary. The normal present-day procedure is to ventilate the sewerage system through the numerous house drains connecting to it. Surface water pipes and drains must not, however, be used for the ventilation of foul drains and sewers. The ventilation of drains is referred to in Building Regulations, 1976, while Clause 40(2) of the Public Health Act, 1936, states that 'the soil pipe from every water-closet shall be properly ventilated'.

Ventilation pipes may be required at manholes where these take the discharge from pumping stations, and also on sewers that can become tide-locked. Vent columns should be carried up independently of buildings to a height at least equal to that of the ridges of nearby houses. These columns are normally specially manufactured for the purpose and are built up of 150-mm or 225-mm cast-iron tubes set on a concrete base. The connection from the manhole will normally be at about 600 to 1000 mm below the surface.

SULPHATES AND SULPHIDES

The generation of hydrogen sulphide is always taking place in sewage. Factors affecting the rate of generation include the temperature and strength of the sewage, the pH value, the age of the sewage, and the velocity of flow. Sulphates may be discharged to the sewers in industrial wastes or they may be present in the surrounding ground; these can directly attack the structures in sewers and at the sewage treatment works.

The effect of sulphates on concrete is discussed in 'Notes on Water Pollution No. 6' [20].

This publication states that:

> there is at present no economically feasible method of coating concrete pipes against sulphate attack. It is necessary therefore to regulate the concentration of sulphates in sewage carried by ordinary Portland cement concrete pipes. Effluent containing a saturated solution of calcium sulphate up to 1500 parts SO_3 per million can be considered as safe, since there will generally be enough dilution to reduce the concentration, under conditions of normal flow, below 1000 parts per million.

If the concentration of sulphates is higher for any appreciable time (up to, say, 5000 parts per million), the pipes should be manufactured from sulphate-resisting Portland cement, provided that the pH value is neutral or alkaline. Supersulphate cement may be used for high concentrations in an acid sewage. If the pipe joints are of the rigid cement mortar type, the cement used in the joints must be the same as that used in the manufacture of the pipes.

Hydrogen sulphide can be produced in sewers from the breakdown of organic sulphur compounds. The sulphur is present in sewage as a constituent of proteins and is also derived from household detergents (see 'Notes on Water Pollution No. 32' [25]). Sulphides are also produced by the reduction of sulphates by bacteria. The production of hydrogen sulphide (H_2S) is greater in hot climates, in sewers which have become damaged by subsidence, and in pumped systems where the sewage may be retained at pumping stations over long periods and then pumped through long rising mains out of contact with the atmosphere.

The hydrogen sulphide gas escapes to the surface of the sewage in gravity sewers and at manholes and tanks. The gas is absorbed into moisture on the surface and is then oxidized to sulphuric acid. The gas itself has an offensive odour; it is poisonous and is therefore a danger to men working in the sewers. The sulphuric acid attacks concrete and metal structures, and this can result in extreme cases in the collapse of a sewer. 'The risk of serious damage to a sewer within 30 years is thought by some authorities to be slight if the concentration of H_2S is below 0·1 mg/l' (see 'Notes on Water Pollution No. 32' [25]).

To prevent the formation of hydrogen sulphide, sewers should have adequate self-cleansing velocities; pumping stations should be designed to avoid long periods of storage of sewage, either in the wet well or in the rising main; and manholes taking the discharge from rising mains should be well ventilated.

TELEVISION SURVEYS

During recent years closed-circuit television has been used by many industries. With specially manufactured cameras and other equipment, it is possible to use TV for the inspection of newly constructed sewers, and also to locate faults in existing sewers. A number of specialist firms can now offer this service.

A camera unit, complete with lamp and mounted on a skid or frame, is pulled through the length of sewer, and pictures are transmitted to a screen mounted in a van, which acts as a control room (see Fig. 11.4). Operation of the winch (and therefore movement of the camera) can be controlled from the control room so that selected parts of the sewer can be studied in detail, and if required photographic records can be made of the pictures. The multi-core cable supplying the current to

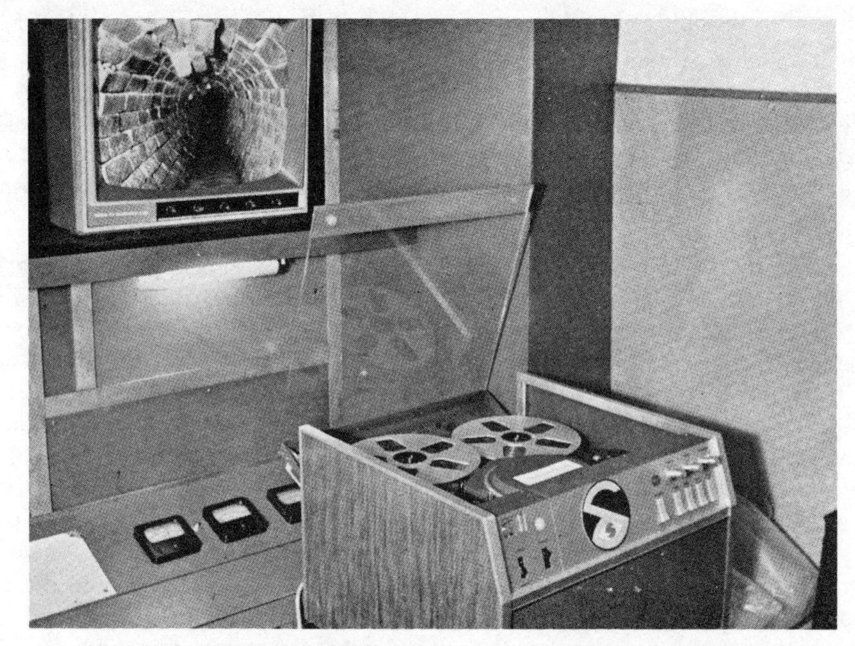

Fig. 11.4. Television survey of a sewer (by courtesy of Seer T.V. Surveys Ltd).

the lamp and transmitting the pictures from the camera is marked so that the distance of the camera from the entry manhole can be ascertained at any time.

An essential preliminary to a television survey is a thorough clean-out of the lengths of sewer to be inspected, as far as this is possible. A nylon cord is then passed through the sewer as a lead to pull through the draw-cord which will control the camera. The nylon cord may be floated down the sewer from manhole to manhole, or it may be necessary to feed it through with drain rods. Valuable time can also be saved later by making arrangements for the power supply, arranging traffic control, and also making arrangements for stopping (or at least reducing) the flow in the sewers during the period of the survey. While the camera can travel underwater without any damage, no picture will be transmitted to the screen if the lens is submerged.

It has been found that if as much as possible of the preparatory work is carried out before the television equipment arrives on site, it should be possible to carry out a thorough inspection of at least 300 m of sewer each day.

12 Sewer Construction

ALTHOUGH sewerage work is occasionally carried out by direct labour, this type of work will generally be carried out under a 'fixed-price' contract. 'Cost-plus' contracts are rarely negotiated for sewer construction schemes. A fixed-price contract may be a 'lump-sum' contract or it may be based on a schedule of agreed rates. More generally, however, a fixed-price contract is based on detailed bills of quantities against each item of which the contractor has quoted a unit rate or price. Payment is then made on the actual quantities of work carried out, these being agreed during the execution of the contract.

Contracts based on open tender are common, but for this specialized type of work it may be more satisfactory to prepare a 'selected list' of contractors during the early stages of design, and to invite tenders only from the contractors included in that list. This ensures that all contractors tendering should have had suitable experience of the type of work. Ministry of Housing and Local Government Circular No. 35/69 'strongly recommends' that local authorities should adopt selective tendering for sewerage schemes (*inter alia*). That circular also draws attention to the need to allow adequate time for contractors to price tenders, and recommends a minimum of four weeks, with up to ten weeks for larger schemes.

Before any contractor's personnel or equipment arrive on the site, it is wise to establish personal contact with each landowner or occupier who is affected by the scheme. so that points of access, office sites, pipe storage, working space, etc., can be agreed. The programme of work can usually be adjusted to avoid unnecessary damage to growing crops, and a basis of assessment of any damages can be agreed. Where relevant, a record should be made of the condition of the land and fences before any work starts. This record should be agreed with each owner or occupier.

It is usual for mechanical equipment to be employed for the majority of trench excavation, hand excavation being restricted to comparatively short lengths where access is difficult or where there are other services. When mechanical equipment is used, a minimum working width of 12 m should be agreed and clearly marked out on site. A wider working width will be necessary for larger-diameter pipelines. The 'employer' under the contract (usually the local authority) will be responsible for damage and surface reinstatement within this working width, while any damage outside the agreed limits will be the responsibility of the contractor.

Turfs should be carefully cut and lifted and retained for re-use later. Where appropriate, topsoil should be removed to a minimum depth of 300 mm and stacked separately. The extent of the removal of topsoil will depend on the site; it may be confined to the width of the trench, but sometimes it is advisable to extend this to include the strip of land to be used for stacking excavated materials.

Where farm stock will use the fields traversed by the pipeline, it may be necessary to allow for temporary fencing along each side of the working width. The minimum requirement is usually the provision of temporary fencing or gates at gaps made in hedges or fences, with provision in the specification for their permanent reinstatement in due course.

Where excavation is to be carried out along or across a public road, the highway authority should be consulted early to agree the line and, where necessary, to make arrangements for road closure or any special traffic control. The construction of sewers in built-up areas presents problems quite different from those encountered in open country. Costs are much higher for such items as road-breaking and reinstatement, traffic control, avoidance or re-routing of other services, and the protection of adjacent properties. Other problems to be considered when working in towns include the effects of noise or vibration, and any special local requirements of traffic or business interests, particularly to avoid claims of financial losses by third parties.

In built-up areas it may be more convenient to remove the spoil from the site and to return it for backfilling later. Precautions must be taken to avoid the blockage of existing surface water sewers; the filling of gully pots with straw or similar material will prevent this to some extent.

LOCATION AND SETTING OUT

Except for the larger diameters, sewers must run in straight lines between manholes. The location of a sewer cannot therefore generally be planned in relation to the centre line of a road. Similarly, depth is governed by hydraulics, and it is usually not possible to arrange the depth of the pipeline in relation to other services.

It is generally considered that such services as gas, electricity, water and telephone should be located under pavements wherever possible. As these pavements usually also house lamp standards and other street furniture, it is often more satisfactory to lay sewers under the carriageway. This will be a very good reason for the inclusion of adequate junctions and laterals at the time of construction, to cater for possible future requirements without having to break open the road surface. It also emphasizes the importance of testing to ensure watertightness at the time of laying.

Wherever possible, the depth of a sewer should be sufficient to allow connections to it to pass under other services. As water mains and high-tension electricity cables are normally laid with a minimum *cover* of 1 m, sewers laid in roads should have a minimum depth to invert of 1·6 m to allow for the outside diameters of both water main and sewer lateral, together with a minimum clearance of 150 mm between the two pipes. In some circumstances the duplication of sewers may be justified where this will prevent interference with traffic or damage to paved surfaces in important main roads.

Particular care is necessary to avoid other services during the excavation of trenches, and prior consultation with the parties concerned is always valuable. This is most important if tunnelling or thrust-boring is to be employed. A telephone call on 'Freefone 111' will provide up-to-date information on the location of post office installations. It is not usually possible to avoid damage to land drains, and information regarding their probable location should be obtained before work starts, so that agreement can be reached with the landowner or occupier as to the best method for their reinstatement.

When sewers are to cross under ditches, culverts or small streams there should be a minimum cover of 450 mm between the top of the sewer and the firm bottom of the ditch. The sewer itself should be protected at the sides and top with 150-mm thickness of concrete. Where a sewer must cross a ditch or stream above the bed, but below the level of the banks, there must be no possibility of obstruction to flow in the stream. Notice of any proposal to cross a watercourse will usually be required under Section 15(2) of the Public Health Act, 1936.

When crossings are to be made under larger rivers or canals, or under railways, special precautions will usually be required by the river or other authority. Section 333 of the Public Health Act, 1936, will probably apply, and it will be necessary to submit details of the proposed crossing for approval before any work is commenced.

Once the general lines of the sewers have been established on site, the centre lines of manholes should be accurately located (spaced to avoid the need to cut pipes if possible), and the centre line of the sewer and the width of trench can then be set out. Levels will normally be based on Ordnance Survey bench-marks, and it is wise to establish a number of temporary bench-marks at suitable points near the route of the sewer. These TBMs can either be on existing structures or they can be on pegs set in the ground, and surrounded with concrete to prevent accidental damage.

Sight rails can then be erected at each manhole location. CP 2005 recommends that:

> a strong sight rail, planed true, painted in contrasting colours, and fixed to rigid posts, should be erected across the centre of each manhole. The centre line of the sewer should be indicated on the sight rail, and its height related in an even dimension to the invert of the sewer. As a check on accidental displacement of a sight rail there should always be at least three rails on each gradient. The length of travellers (fitted with projections to rest in inverts) used between sight rails should be checked at frequent intervals.

If a string is stretched along the trench at mid-diameter of the pipes and just clear of the collars it is possible to ensure that the pipes are laid in a straight line.

During recent years the use of sight rails and travellers has been superseded, to some extent, by the introduction of the laser beams. The advantage of using a laser is that it produces a stable pencil-thin beam of light which is detectable even in daylight conditions at a range of up to 400 m or more. Once set on line, this can be used to control the work as it progresses, eliminating the necessity of a second man at the instrument (as is the case with a theodolite or level). The electrically operated instrument is usually set up at one manhole location and once set to line and level, pipelaying is a fairly simple matter of lining up the pipe using the 'target'. Many laser beam units are 'self-levelling'.

PIPELINES IN TRENCH

It is generally advisable to allow the contractor to choose his own method of excavation, provided that it complies with the specification and drawings. The pipeline will, however, have been designed for a specific maximum trench width (see Chapter 10), and it must be clearly set out in the contract documents that any proposals by the contractor to excavate beyond the design width must be approved by the engineer. The cost of any additional work which may then be required to maintain the strength of the pipeline must be borne by the contractor.

Trenches should normally be dug to a minimum width of 300 mm, plus the diameter of the pipeline where this is 150 mm or more. Additional width should be included for any timbering or sheeting. A formula used in America provides for a trench width of $(1{\cdot}33d + 200)$ mm, where d is the normal diameter in millimetres. A similar formula proposed by the Clay Pipe Development Association gives a trench width of $(1{\cdot}67d + 250)$ mm, within a minimum width of 500 mm.

Whenever possible, the width of trench below the level of the crown of a sewer should be restricted to that required for proper laying and jointing of the pipes. Above the pipes it may be more economical to have sloping trench sides to avoid timbering, but this must depend on site conditions. Excavation of trenches with sloping sides should always be subject to the approval of the engineer.

In the event of a slip in the sides of the trench below the crown of the pipes, the width of the trench must be considered as the width measured to the undisturbed soil. This width must be compared with the structural design, as it may be necessary to strengthen the bedding to withstand any increase in loading on the pipeline.

CP 2005 refers to the construction of sewers in made ground, peat and coarse sands where these are waterlogged, and points out that:

> dewatering of these and silts can be difficult, dangerous and very costly, it may cause the subsidence of existing structures or underground works by loss of water or fines, or both, and it should usually be carried out under expert advice. Under such conditions trenching may prove impossible and thrust-boring may be quicker and less expensive; it may also be safer in the vicinity of existing structures since it causes less disturbance of the ground.

Excavation in rock or other firm strata may not require any support, while, on the other hand, excavation in soft, fine materials may warrant the use of close sheeting. The type of support to be provided to trench sides will vary from simple poling boards, held apart by struts across the trench, to timber runners taken down as the trench is excavated. Alternatively, steel sheet piling may be driven down along the line of the trench before excavation is begun. This latter method is, of course, the most expensive, but it has the advantage of reducing the rate of infiltration of subsoil water. Attempts to economize on construction by using inadequate supports to the trench sides can provide dangerous working conditions, and may result in cracks in adjacent paved surfaces which are costly to reinstate.

Timber sizes will vary with trench depth and the type of subsoil. Poling boards should be about 150 by 50 mm minimum, while horizontal walings should be at least 100 by 75 mm. Timber struts can be from 75 by 75 mm to 300 by 300 mm, depending on the depth and width of the excavation. Close sheeting in timber can generally be carried out in 225 by 38 mm timbers, in lengths up to 4·2 m, according to trench depth.

Acrow and similar adjustable trench struts are available in various sizes for trench widths (inside walings) from 300 to 1700 mm. Patent frame-type trench shores are also available for trenches up to about 1300 mm in width.

Light steel trench sheeting is available in lengths up to about 4 m. This is more easily driven than timber and can be withdrawn and re-used. The overlapping-type sheets normally used for temporary supports to trenches have an effective thickness of 35 mm, are 330 mm wide (centre to centre) and are usually driven in groups of up to seven or eight.

Where trench struts are used above or under a pipeline, and where they are likely to be left in position after backfilling, they should be kept well clear of the pipes, so that they do not constitute hard spots in the foundation or backfill.

Mechanical excavation of trenches with trenching machines (ladder or wheel type) was popular some years ago, but most contractors now prefer to use either backacters (back hoes) or dragshovels. In very deep excavations, clamshell buckets may be used to advantage, or a crane and open bucket may be employed. Trenchers can work to depths of about 4 m and, as they are particularly useful for very narrow trenches, they find more favour for the laying of smaller diameter water mains. Backacters can work to depths of about 6 m. Dragshovels are particularly useful in open country (e.g. excavations with sloping sides) and can operate at depths down to 11 m.

Trenches should not be opened too far in advance of pipelaying, and they should be backfilled as early as possible after the sewers have been laid and tested. When granular bedding material is used, it should normally be possible to carefully withdraw any trench sheeting as the backfilling proceeds. If support is required for adjacent structures, sheeting may be cut off below ground level and left in position.

Excavation should commence at the lower end of a pipeline and should proceed upstream. This will allow pipe-jointing to be carried out with the sockets facing upstream, and will also permit subsoil water to drain away from the working area. If necessary, temporary drains should be laid in the bottom of the excavation, leading to a sump at the lower end. Should the excavation be taken down below the required depth, the extra depth must be filled with compacted granular material (where pipes are to be laid on a granular bed), or with concrete if a concrete bed or surround is to be used.

The laying of pipes directly on the natural trench bottom should be restricted to pipes with flexible joints of 300-mm diameter and under (except when cast-iron pipes are used), where dry conditions can be achieved, and where the subsoil is such that accurate hand trimming is practicable. Where this form of bedding is to be used, and where mechanical equipment is used for excavation, the trench should initially be underdug, and the last 100 mm or so should be carefully excavated and trimmed by hand, together with the socket holes, so that the pipes will be supported along the full length of their barrels.

HEADINGS, TUNNELS AND THRUSTBORES

When a sewer has to be laid under a main road, railway, canal or similar obstruction, it may be advantageous to avoid open cut and to lay the sewer in heading or tunnel, or to use the thrust-boring techniques which have been developed recently. Work in heading or tunnel can be carried out throughout the twenty-four-hour day more or less irrespective of the weather. Other than in very exceptional circumstances, it is usually less expensive to lay sewers in open cut for depths down to about 6 m.

A heading is possible in firm, cohesive soil or in rock. CP 2005 recommends that the smallest size of heading for proper working is about 1140 mm clear height by about 760 mm width at the bottom. It is often possible to employ a combination of open trench and heading when laying sewers along urban roads. For smaller-diameter sewers in heading, the longitudinal timbers should be about 225 by 75 mm or 250 by 100 mm, while framing members should be 250 by 250 mm.

In all cases of work in heading, great care is needed with the backfilling and consolidation around the pipes. As it is usually necessary to do this as pipelaying proceeds, this is simplified by the use of flexibly jointed pipes set on a granular bedding. The timbering must often, of necessity, remain in position to avoid any risk of collapse, but it should be withdrawn if possible to reduce the possibility of hard spots in the bedding. Weak concrete is recommended as a backfilling material for shallow headings, although sand and other readily compactible materials have been used successfully, particularly in deeper headings where a small gap at the top of the packing will not have any serious detrimental effect.

Traditional tunnelling methods, in which the tunnel is formed approximately to the required slope of the finished sewer, may be used for larger-diameter sewers. For sewers of 1400-mm diameter and over, tunnels can be of cast iron or concrete segmental rings, lined with concrete or brick.

Thrustboring can now be used for pipes of diameters up to 2600 mm. Reinforced concrete pipes of diameters from 900 to 2600 mm are manufactured with special square-shouldered flexible joints which are suitable for jacking. There are problems in maintaining true line and level, particularly if the subsoil is variable, and the cost of the thrust pits and the blocks required for the jacks can make this method expensive. It is, however, very suitable for making crossings under railways and important highways, and in certain circumstances there can be a saving in cost, in addition to a considerable reduction in disruption to traffic flow. The Concrete Pipe Association has issued a Technical Bulletin on the Jacking of Concrete Pipes [90] in which are set out recommended methods of jacking, together with a typical specification and bill of quantities.

DEWATERING

The quantity of water to be removed from the trenches and other excavations must be estimated in the early stages of construction so that suitable construction methods can be adopted. In many cases, the drainage of trenches to one or more sumps may be sufficient, the water level in the sumps being kept down by pumping more or less continually, using standard contractors' pumps.

In suitable soil conditions, when the inflow of water is too great for normal pumping, well-point dewatering can be satisfactory, but this method of lowering the ground water 'cannot be applied to soils containing more than 10 per cent by weight of grains less than 0·03 mm diameter. Silts, in which most of the particles are less than 0·06 mm diameter, cannot be successfully dealt with in this way' (CP 2003).

The well-point system consists of a series of vertical 'well-point' pipes and risers (approximately 40- or 50-mm diameter), sunk into the water-bearing stratum. These are connected through short horizontal pipes and valves to a horizontal header pipe (usually 150-mm), which is in turn connected to a vacuum pump. While one line of header pipe and risers may be sufficient, it is usually necessary to sink risers on both sides of the line of the proposed trench. The well-points are sunk into the ground by 'jetting', i.e. by forcing water through them to scour away the ground beneath the well-point. In some circumstances the normal well-point vacuum pump can be used (in reverse), but where the riser must penetrate hard strata a special multi-stage centrifugal jetting pump will be required. After the risers and head pipes have been fixed and connected up, the pump is used to form a vacuum in the system. This allows the ground water to rise in the risers so that it can be pumped to waste clear of the excavation.

When excavation is to be carried out to depths below normal pump suction limits, it may be necessary to carry out the first 5 m or so of excavation, and then to set new well-points, complete with risers, header and pump at this lower level, before any further excavation is possible.

PIPELAYING

Each pipe should be examined carefully on delivery and any that have been damaged must be clearly marked and removed off the site. Factory-applied joints on pipes should be protected according to the manufacturer's instructions. All pipes and joints should be examined again immediately before they are laid.

Where the pipes are to be laid directly on the trench bottom, this should be trimmed to correct level and gradient immediately before the pipes are laid. The trimming can be checked by 'boning in' a traveller over the sight rails. Socket holes should be formed at each socket position, leaving the maximum length of support for the pipe barrels. To ensure that the pipes are laid to the correct gradient each pipe is then set individually to line and level, using the sight rails, or laser.

If the pipes are to be set on a granular bed, this must be laid first to an approximate level and gradient. Socket holes are then scooped out and each pipe is bedded into the granular material using the sight rails or laser to obtain correct line and gradient as before.

Where concrete protection is to be used (other than a concrete arch), it may sometimes be convenient to put down a weak concrete mat (about 50-mm thickness) to form a clean working surface. This mat should be set to approximate level and gradient. The method of placing the concrete for cradle, haunch or surround varies among engineers. Some prefer to lay a fairly dry concrete bed (with socket holes) and then to complete the concrete after the pipes have been laid, jointed and tested. Others prefer to lay the pipes on precast blocks set on the trench bottom (or the mat concrete), and after laying and testing the pipes the whole of the concrete is then poured in one operation. Both methods have their disadvantages: with two-stage concreting it is difficult to obtain a good bond between the two layers of concrete, while with one-stage work, it is difficult to get adequate concrete support immediately under the pipes.

Pipelaying should normally start at the lower end of a line, working with sockets pointing uphill. The 'traditional' rigid joints of vitrified clay and concrete pipes are made with a tarred hemp gasket, caulked tightly home so that it fills no more than a quarter depth of the socket, together with a stiff cement/sand mortar to fill the remainder of the space in the socket; this is usually finished off with a 45° fillet. Mortar mixes vary from 1:1 to 1:3, depending on the jointer's personal preference. Rich mixes can cause cracking and leakage at the joints, but the mix must be strong enough to adhere to the pipes and to withstand the test pressure without undue sweating. As the contractor must accept responsibility for the workmanship, it is preferable not to include any mortar mix in the specification. The insides of the pipes must be cleaned with a damp cloth, and a close-fitting pad should be drawn through each pipe as it is laid.

Flexible/mechanical joints are available with pipes of most materials. These joints should be used in preference to rigid joints whenever possible. The majority of the failures in drains and sewers in the past have been caused by, or aggravated by, the use of rigid cement mortar joints. Jointing pipes with mechanical joints should be in accordance with the manufacturer's instructions, which will vary to some extent between different pipe materials and different diameters. The annular space outside the jointing ring must not be filled with mortar, as this will impede the free flexible action of the joint and may cause burst sockets. Where there is a possibility of this annular space becoming filled with gravel or stones, it can be filled with puddled clay or with fine soil.

If mechanical joints have been used specifically to obtain flexibility in the pipeline (e.g. in areas subject to mining subsidence, or through soils which are liable to swell and shrink), any concrete protection must also be able to move in unison. Joints must therefore be incorporated in the concrete (usually at pipe joints), and before any concrete is placed the annular spaces at pipe joints should be protected to prevent the intrusion of concrete. This can be accomplished with clay or hessian.

While small-diameter pipes with flexible joints should be jointed manually, this is not possible with larger diameters. Mechanical pulling devices will then be needed, and care should be taken to follow the manufacturer's instructions. Small-diameter pipes can be trued to line and level after jointing, but larger pipes are difficult to lift and must usually be laid in their final positions as they are jointed.

Fig. 12.1. Trenchless pipelaying (by courtesy of Hudswell Badger Ltd).

The procedure for laying most flexible pipes (e.g. PVC and pitch-fibre) is similar to that for rigid pipes on a granular bed, except that more attention must be given to the side support, as this is an essential parts of the strength of the completed pipeline. After the pipes have been laid to correct line and level and the test has been completed, it is usual to place and thoroughly compact selected bedding material up to a level above the crown of the pipe, and between the pipe and the undisturbed soil of the trench sides.

Flexible pipes should not normally be bedded on concrete. If the depth of cover will be less than about 0·5 m the pipes should be protected from damage by pickaxes, etc. It should, however, be appreciated that laying pipes at this shallow depth is generally bad practice.

MOLE PLOUGHING

Mole ploughing has been used for many years as a method of laying agricultural drains and for small-diameter flexible pipelines—particularly for rural water supplies. As this method eliminates the need for trench excavation, it is usually quick and economical, but it is, of course, not suitable for all types of soil, nor can it be used in ground containing other service pipes or cables.

When using solvent welded PVC pipes, the pipes should be jointed the day before laying. The pipeline can then either be laid using a modification of the original land-drain mole plough pulled by a tractor (directly or by cable with the tractor anchored), or by a specially manufactured trenchless pipelaying machine which pulls the PVC pipe into the duct behind the blade of the machine, generally in a single operation.

Trenchless pipelaying methods have been developed recently and can now be used for laying pipes to specified gradients. With the more sophisticated machinery now available, PVC pipes can be laid as gravity sewers. Machines have been developed which will lay pipes up to 375-mm diameter to depths of about 3 m with accuracy of both line and level. One of these machines is illustrated in Fig. 12.1. Sensitive grading, steering and uprightness of the blade are achieved by using a 'floating blade' controlled either manually or automatically, using a beam of infra-red light. Pipelines can be laid to gradient with an accuracy of plus or minus 12 mm vertically and 75 mm laterally.

BACKFILLING

Backfilling under roads is a structural operation, and should be treated as such. A lightly hand-tamped layer of selected material, preferably granular and not containing any large stones or other hard objects, must be laid immediately above any type of pipe to act as a shock-absorber for the compacting operations in the fill above and to relieve some of the load on the pipeline. Above this level, compaction should aim at the same density and moisture content as those of the undisturbed soil in the trench sides, so as to avoid subsequent settlement and consequent disturbance to the road surface and to any underground works in or near the trench. If the latter works are supported less or more by the backfill than by the undisturbed soil into which they pass, they may be fractured. It is not sufficient just to push the soil back into the trench; in deep trenches such careless filling may even cause overload fractures of the pipes.

The initial lightly tamped layer should extend to 300 mm above the crown of the pipes. This layer must not contain large stones, roots or lumps of clay or frozen soil. Once the remainder of the fill has been placed in position and adequately tamped, road reinstatement should be carried out without delay.

Backfilling over *flexible* pipelines must be carried out with extra care to ensure that the side fill affords the necessary support to the pipes. The material for filling, at both the sides and over the pipes for at least 100 mm depth, should be similar to the bedding material.

Subsequent filling of trenches and around manholes should be built up in layers not exceeding 150 to 230 mm (uncompacted thickness), and each layer should be thoroughly compacted before any further material is added. The intention is to have a completed job as nearly as possible equivalent to the undisturbed soil. While some highway authorities may ask for the filling to be of weak concrete, this is most undesirable, as it can be responsible for subsequent damage to both the pipeline and the road itself.

When pipelines cross roads, railways, rivers and canals, or where the pipeline crosses open country, it is useful to erect distinctive marker posts at field boundaries, etc. These will not normally be required for gravity sewers where the manhole covers are left visible, but they should be used for rising mains.

TRAFFIC CONTROL AND SAFETY ON SITE

With the increase in density and speed of traffic, the necessity for proper warning signs and traffic control arrangements on any public or private road has become more important. Details of all relevant and authorized traffic signs are contained in the Traffic Signs Regulations, 1964, and in the appropriate Traffic Safety Code for Road Works.

The minimum standard of warning on any excavation in a carriage-way should be a 'Road Works' sign (No. 564), facing in each direction in two-way traffic roads, together with traffic cones around the working area. These should be supplemented by 'Road Clear' notices (No. 829) to inform drivers when they are clear of the obstruction. Suitable distances ahead of the obstruction for sign No. 564 are:

on 100 km/h roads . . .	140 m
on 65 km/h roads . . .	70 m
on 50 km/h roads . . .	35 m

On very fast roads these signs should be repeated at 1 km and, if necessary, at 2 km before the obstruction (along with sign No. 571).

Additional signs, such as those for single-file traffic (No. 518) and priority (No. 615), should also be used as necessary, while traffic control with 'Stop-Go' boards or automatic traffic lights is now almost essential except in very wide roads.

If the warning signs are to be effective it is essential that they are moved regularly as work on trench excavation or reinstatement proceeds along a road. It is equally important that signs no longer relevant should be removed from sight as early as possible.

At night-time and during times of poor visibility all excavations must be adequately lit. The newer flashing-type light beacons are far superior to the older type hurricane lamps and are referred to in Clause 28 of the 1964 Regulations.

The warning signs referred to in the 1964 Regulations are mainly for the benefit of vehicle drivers. It is, of course, equally important to safeguard pedestrians. Excavation in pavements must be properly fenced and lit, and temporary pedestrian ways should be provided where possible. Special precautions necessary for the protection of the blind are referred to in Ministry of Transport Circular No. 17/67. A light fence about 600 mm high is used by the Post Office to guard all excavations in footways, and this type of guard should be adopted wherever possible.

The safety of the workmen on site is governed by the various Construction Regulations, 1961 and 1966, and to some extent by the Factories Act. Of particular interest on civil engineering contracts are the precautions necessary to ensure safety on scaffolding and ladders, and the use of guard rails or covers at openings in floors and roofs. Guard rails and toe boards must be provided around working platforms on scaffolding. Ladders must be properly secured and must rise to at least 1 m above the place of landing. Safety helmets are recommended for use where work is proceeding at two levels on a site, while safety harness should be used by men working at heights. Every construction site should have a first-aid box, which must be readily accessible in an emergency.

BSCP 2003 lists a number of sources of danger and the expert help or advice which should be sought before proceeding with work if these are found or are to be expected during excavations. These are as follows:

1. Noxious gases, e.g.

Carbon monoxide Sulphur compounds Carbon dioxide Air lacking oxygen Town gas Methane	H.M. Inspector of Factories H.M. Inspector of Mines Public Analyst Local Gas Board Medical Officer of Health

2. Noxious or inflammable liquids, e.g.

Petrol Alcohol Cleansing chemicals White spirit Acids Alkalis	H.M. Inspector of Factories Medical Officer of Health

3. Unexploded missiles Police

Safety measures which must be taken when working *inside* sewers (particularly sewers in use) are very distinct from those required on other types of work, due particularly to the danger from poisonous gases, bacterial infection and flooding. Methods of eliminating risks, the precautions to be taken, and the equipment required are set out very fully in the Institution of Civil Engineers publication *Safety in Sewers and at Sewage Works* [36].

Probably the most important precautions to be taken when operating in existing sewers are:

1. The sewer must be well ventilated *before* it is entered by workmen. The minimum requirement is the removal of manhole covers upstream and downstream of the section involved.

2. No one should enter or travel along a sewer alone. Two top-men should be present at any manhole from which work is in progress.
3. Naked lights must not be used in a sewer or within 3 m of an open manhole.
4. The nose, mouth and eyes should not be rubbed with dirty hands. After contact with sewage, the hands and forearms should be thoroughly washed before taking any food or drink. Any cuts or abrasions should be washed and dressed as early as possible.

13 Testing Sewers

SEWERS and drains are tested after laying to ensure that they are sound, i.e. that no damaged pipes have been laid and that the joints are satisfactory. If the construction of the sewers complies with BS codes of practice and the Building Regulations, the pipes themselves (and the joints where applicable) will be to a BS or equivalent specification, and will normally have been tested before leaving the manufacturer's works.

It is important that any test should be realistic, relatively easy to apply and reliable. It should not be possible to pass a faulty line nor to reject a line which would give satisfactory service. Specifications should be so written that only requirements are included which are expected to be attained and which it is intended to enforce, and clauses which lay down absolute requirements which both the engineer and the contractor know to be unattainable in practice should be excluded.

BSCP 2005 recommends that tests for watertightness should be made on all lengths of sewers and drains up to at least 760-mm diameter, together with all manholes and inspection chambers. Regulation N11 of the Building Regulations, 1976, stresses the need for testing *after* backfilling the trenches.

In the past, testing for watertightness was generally achieved by means of a water test, but if the full benefits of using mechanical joints are to be obtained, a quick and simple form of test is essential.

On many sites it may also be difficult or expensive to obtain water in sufficient quantity for a water test, and as a result many sewers are now tested with an air test. The CP air test is, however, not satisfactory for testing inverted siphons or rising mains, nor can it be used as a test after backfilling. Tests before backfilling should reveal faulty pipes and joints, while those carried out afterwards may disclose any faults in bedding or any subsequent damage during or after backfilling.

Wherever possible, testing should be carried out from manhole to manhole. Short branch drains connected to a main sewer between manholes should be tested as one system with the main sewer. Long branches and manholes should be tested separately (CP 2005).

Tests with smoke and similar substances are not suitable for newly constructed drains and sewers, and should be used only for the detection of faults on existing drains. Tests for straightness and lack of obstruction and checks on infiltration should be carried out after backfilling.

THE BUILDING REGULATIONS AND CODES OF PRACTICE

Regulation A14 of the Building Regulations, 1976, permits a duly authorized officer of the local authority to make such tests of any drain or sewer as may be necessary to establish compliance with the provisions of Part N of the Regulations. Provision for carrying out a suitable test for watertightness, after laying and backfilling, is included in Regulation N11.

To locate any faulty pipes or joints, a pipeline should be tested before any concrete protection is placed. Further tests after backfilling may reveal faults in bedding or any subsequent accidental damage. The BS codes of practice assume a water test as the normal test, with an air test as an alternative. Failure to pass an air test is not to be taken as a reason for the rejection of a pipeline, and in the event of such a failure a water test should be made to determine the leakage rate before any decision is made.

The codes do not include a watertightness test for large-diameter pipelines. When pipes are large enough to be entered (about 1000-mm diameter and over) reliance is usually placed on a physical inspection of the pipes and joints as work proceeds, together with a final inspection from the inside after backfilling has been completed.

INFILTRATION

Excessive infiltration of subsoil water into drains and sewers decreases their effective capacity, and increases the cost of pumping or treating the sewage. It is therefore desirable to keep infiltration to a reasonable minimum. It will be highest in waterbearing ground where the water table is high, and for that reason it is usually higher in winter than in summer. Infiltration may be the result of damage caused by mining subsidence.

Many years ago Bevan and Rees recommended that infiltration into concrete and clayware sewers should not exceed about 0·002 m^3/min/km. This is equivalent to about 12 litres/h per 100 m of sewer. The loss allowance during a water test on a 150-mm diameter pipeline under CP 2005 is equivalent to 15 litres/h per 100 m.

The introduction of mechanical/flexible joints with both concrete and vitrified clay sewers, and the emphasis now placed on the structural design of pipelines and their bedding, have substantially reduced the possibility of infiltration into new sewers. However, where old drains or sewers are to be connected into a new system (particularly in waterlogged ground) it is wise to make a substantial allowance for infiltration from those drains. The actual amount to be allowed can be assessed from observations of flows at night, or by the comparison of actual flows with calculated dry-weather flows.

THE CHOICE OF TEST FOR A NEW SEWER

As the pipeline has eventually to carry a liquid, the most logical and satisfactory test is a water test. This measures the rate of exfiltration from the sewer and therefore gives a direct measurement of how much water will pass out of the pipeline under given conditions. For all practical purposes, the rate of exfiltration can be regarded as indicative of the relative infiltration to be expected.

The water test, however, has certain disadvantages:

i. It can require large quantities of water.
ii. Time must generally be allowed for absorption by the pipes before the test is commenced.
iii. Air pockets may be formed—resulting in false test readings.
iv. There may be a problem in disposing of the water after the test.

The time taken to carry out a water test means that some of the benefit of using mechanical/flexible joints is lost. It is therefore often more convenient to test drains and sewers with air pressure, as this type of test can be carried out immediately after the pipes have been jointed, and can be completed within a few minutes.

The air test is, however, easily affected by very slight leaks at stoppers and other testing equipment, and is affected to some extent by temperature changes. The air test does not measure the loss of volume of air, but is a measure of the loss in pressure. It therefore does not give an accurate indication of the condition of the pipeline.

THE WATER TEST

When tested with water, all drains and sewers should generally be subjected to an internal test pressure of 1·2 m head of water (about 120 mbar) above the crown of the pipe at the high level, but not more than 6·0 m (600 mbar) at the low end. With a length of steeply graded drain or sewer, if the head at the lower end would exceed 6·0 m, the line should be tested in stages.

The codes of practice set out the procedure for preparing the test, and recommend that, after filling the line, an appropriate period should be allowed for absorption of water by the pipes (a period best to be determined by conferring with the pipe manufacturer).

The loss of water over a period of 30 min is then to be measured by adding water from a measuring vessel at regular 10-min intervals and noting the quantity required to maintain the original level. The average quantity added for pipes up to 460-mm diameter should not exceed 1·0 litres/h per linear metre of pipeline per 1000 mm of nominal internal diameter.

The head of 1·2 m at the top end of the pipeline can conveniently be obtained by temporarily jointing a 90° bend and a straight pipe of the same diameter as the pipeline to the head of the line. Provided that the diameter of the standpipe used is the same as that of the pipeline being tested, the drop in water level in the standpipe for every 100 m of pipeline and *during each* 10 *min of the test* should not exceed:

$$h = \frac{25 \times 10^3}{d} \text{ mm} \qquad \textbf{Formula 13.1}$$

where
h is the drop in the level in the standpipe
and d is the diameter of the pipeline in millimetres

Formula 13.1 is shown graphically in Fig. 13.1.

As a gravity sewer is rarely designed to run full, it will be appreciated that the water test imposes conditions which are normally more severe than are met in practice. If a sewer is designed to work surcharged (e.g. an inverted siphon), it should be constructed with pressure pipes and tested accordingly.

To calculate the approximate amount of water required to fill a pipeline use the following formula:

$$Q = \frac{d^2 L}{1275} \text{ litres} \qquad \textbf{Formula 13.2}$$

where
d is the diameter in millimetres
and L is the length in metres

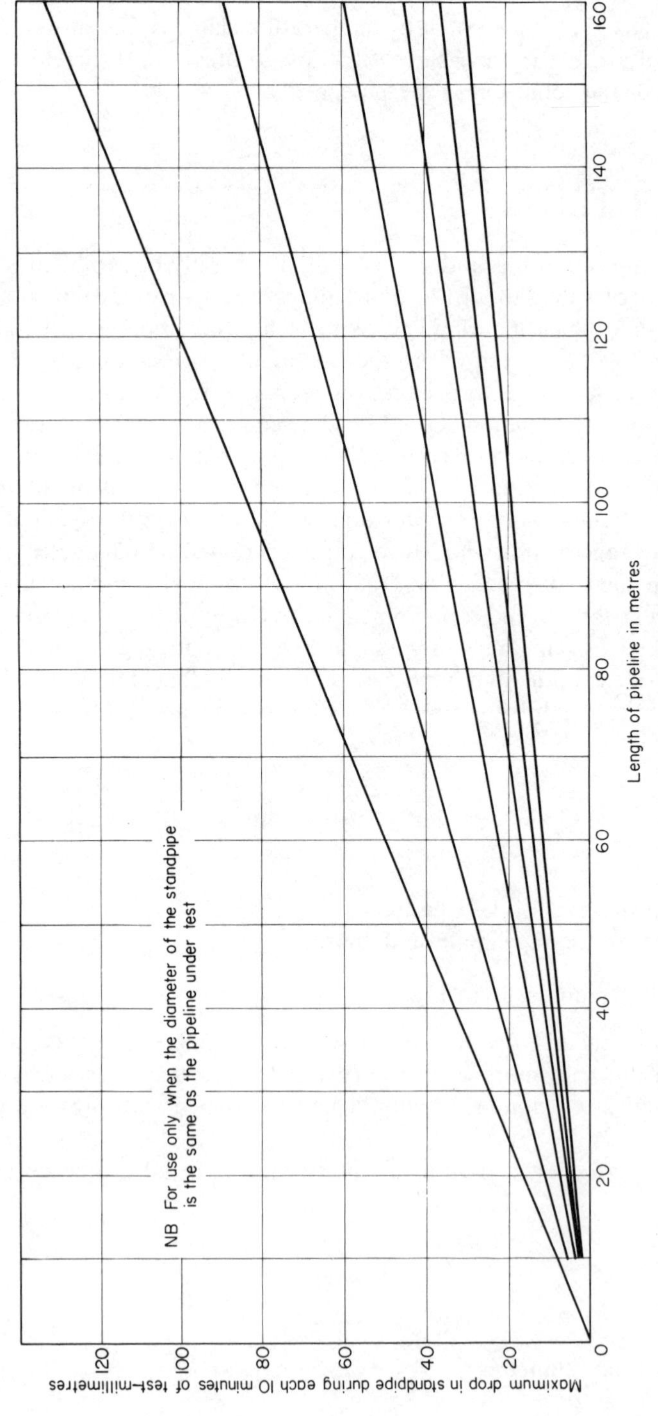

Fig. 13.1. *Code of practice water test—permissible drop in standpipe (see Formula* 13.1).

THE AIR TEST

The air test is now generally considered to be the more convenient method of testing drains and sewers before backfilling. It should not be used for testing *after* backfilling or if the trench is waterlogged, as under those conditions it is possible for a faulty line to appear to be satisfactory.

Care is needed when using the air test, as change in temperature during the test can affect the reading on the manometer, due to the comparatively large expansion of air with a rise in its temperature. For a difference of 1 °C the pressure of the air in the line will alter by 38-mm water gauge.

The requirements of the codes of practice for the air test are that:

> the length of pipe under test should be effectively plugged and air pumped in by suitable means until a pressure of 100 mm of water is indicated in a glass U-tube connected to the system. The air pressure should then not fall to less than 75 mm during a period of five minutes without further pumping, after allowing a suitable time for stabilization of the air temperature.

The test pressure recommended should not be exceeded, as the use of high air pressures entails the storage of considerable potential energy which can be extremely dangerous if released suddenly.

As the loss of air pressure in an air test is not directly related to the surface area of the pipeline, this test is not a direct measure of the rate of loss, but is more an indication of the possibility of a fault in the line. The codes state that 'failure to pass this test is not necessarily conclusive and when failure does occur, a water test should be made'.

DEVELOPMENTS OF THE AIR TEST

In view of the limitations of the air test described above, possible variations of this test have been proposed from time to time. These would, however, generally entail the use of either more complex equipment or of formulae and charts not conveniently used on site.

Workers in New Zealand have proposed the use of a 'control' pipe on site, in an attempt to eliminate the effects of changes in air temperature. This envisages the use of a 'standard pipe' taken from the stack on site, together with two linked manometers. It is doubtful whether the temperature effects are sufficiently eliminated to warrant the use of the more complicated equipment required.

In South Africa, Shaw of the National Building Research Institute [60] has examined the possibility of using a modified form of air test to locate infiltration into sewers, and for the investigation of sewers and drains after they have been in use for some time. The basis of those tests was to establish *the rate of air leakage* at a constant pressure, and for this he produced a formula with constants for various pipe diameters ranging from 80 to 450 mm.

In 1964 a paper on 'Low Pressure Air Test for Sanitary Sewers' [57] was given by Ramseier and Riek to the American Society of Civil Engineers. That paper reported on investigations aimed at detecting what *rate of air loss* was significant when testing sewers. The pressures used (up to 3500 mm water gauge) were much higher than those used in the UK, and the equipment and the procedure for the tests were therefore designed for the higher pressures involved. The essential principle of the Ramseier test is that the rate of air loss is related to the internal surface of the pipeline, and that with any increase in the length of the pipeline there is an increase in the opportunity for leakage due to damaged pipes or faulty joints. Ramseier therefore proposed formulae for the calculation of the *length of time of a test* according to the diameter and length of the pipeline. To facilitate the calculations on site, he produced a nomograph to use with the formulae. The Ramseier formulae and nomograph could not be used with the BSCP air test, in view of the differences in pressure used.

As the codes recommend tests for watertightness for pipe diameters up to 760 mm, a possible simplified parallel to the Ramseier test might be to take a 300-mm or 375-mm diameter pipeline as an *average* diameter for the CP test time of 5 min, and then to adjust the times of test for other diameters *pro rata* up or down on the basis of the surface area of the pipeline. This would help to eliminate the effects of changes in air temperature when testing smaller-diameter sewers; possible times of test for smaller diameter pipelines based on surface area might then be:

150-mm diameter: 2 min
225-mm diameter: 3 min
300-mm diameter: 4 min
375-mm diameter: 5 min (as CP 2005)
450-mm diameter: 6 min
525-mm diameter: 7 min
600-mm diameter: 8 min

OTHER TESTS

Tests using smoke or strong-smelling essences (the olfactory or odour test) may be used for house drainage. As the Public Health Act, 1936, does not permit the use of a water test on existing drains, these tests are useful when locating bad leaks on old pipelines, but they are not satisfactory for new work, and should not be used in lieu of a water or air test.

A smoke test can be applied by the introduction of a smoke rocket (approximately 200-mm long by 40- or 50-mm diameter) into the drain, and then plugging both ends of the line to be tested (e.g. at two inspection chambers). An air pump is then used to increase the pressure in the drain. As an alternative, a 'smoke machine' can be used. This takes the form of a bellows and a chamber for the production of smoke under pressure. Smoke testing of uPVC plastic pipework and fittings is not recommended. There are certain smoke generating devices on the market whose products of combustion are detrimental to plastic pipework. It is essential that any person intending to carry out smoke testing on uPVC pipework must satisfy themselves by contact with the manufacturer of the device that there will be no detrimental effect.

The olfactory test may sometimes be usefully employed for testing soil and waste pipes, either with a chemical drain tester (which is withdrawn from the pipe after use) or by using oil of peppermint or some other strong-smelling substance in solution.

Possibly one of the most useful tests to apply to a newly constructed drain or sewer after backfilling is the 'light test'. This effectively checks the line for straightness and for freedom from obstructions. The tester uses a mirror held at 45° to the axis of the pipeline in one manhole, while his assistant shines a torch from the next manhole. If the line is free from bends and obstructions, there should be a clear image of the light in the mirror.

If the crown of the pipe at the high point of the length of sewer under test is more than 1200 mm below the water table, an infiltration check should be carried out *after* backfilling. All inlets must be effectively closed off, and if 'more than a trickle' is evident in the manholes or inspection chambers the infiltration must be traced and rectified.

Where it is thought that a sewer is obstructed or that there is excessive infiltration, a television survey may be the most economical method of locating the faults. Under quite difficult site conditions, it is usually possible to examine and record details of 300 m or more of sewer per day (see Chapter 11).

TESTING PROCEDURE ON SITE

Before carrying out either a water or an air test all equipment must be carefully checked. It is most important that the stoppers and other equipment used for air testing are not worn or damaged, and that they are free from grit, etc.

The pipeline itself should be inspected to ensure that there are no obvious faults and the ends should be strutted to prevent movement. If any cement mortar joints are incorporated in the line at least twenty-four hours must elapse after jointing before the test is carried out.

If a water test is being carried out, the most satisfactory standpipe to obtain the requisite 1200-mm head consists of a 90° bend and a vertical pipe of the same diameter as the pipe being tested. Filling the line is then conveniently done through the standpipe, and should *never* be done through a hose and drain plug unless there is a standpipe open to the atmosphere.

For an air test, plugs are firmly fixed at both ends of the line, and air is pumped or blown in until the pressure slightly exceeds the test pressure of 100-mm water gauge as indicated on the U-tube of the manometer. After adequate time has been allowed for the temperature of the air in the pipeline to cool down to pipe temperature, the pressure should be adjusted to 100-mm water gauge and the drop noted during the period of the test. The final pressure should not be lower than 75-mm water gauge.

Both the water and air test can be 'rigged' by an unscrupulous contractor. If in any doubt, the resident engineer or clerk of works should take precautions to ensure that no intermediate plugs have been inserted in the line. For an air test it may be preferable to have the manometer at one end of the line and the pump at the other. For a water test it is usually sufficient to witness the release of the water after the test.

14 Pumping Stations

IT is usually preferable for any sewage treatment works to be at the lowest point of a sewerage system, and for all sewage to flow to that point by gravity. This is, however, often not possible, and it may be necessary either to pump the flows from small, isolated sections into neighbouring higher sewers or, where the district is very flat, it may be necessary or more economical to pump the sewage as an alternative to the construction of very deep gravity sewers. It is not economical to lay a very deep trunk sewer merely to serve a few low-lying areas en route.

In some instances the solution will be obvious, but in others the relative economics may have to be studied in detail so that the capital cost of deep sewers, in trench or tunnel, can be compared with the capital and annual costs of pumping. If a pumping station is required, it will normally be at a low point in the system, and care must be taken that the site is not liable to flooding, and that suitable facilities exist for an emergency sewage overflow in the event of power failure or mechanical breakdown.

Although the smell or noise from a well-designed pumping station is generally very limited, there is always the possibility of complaint if a station is constructed near residential property. To avoid any suggestion of nuisance or of interference to radio and television reception, pumping station sites must be chosen very carefully, bearing in mind both present and probable future developments. In any event, the external appearance of the building should blend with the neighbourhood, and suitable precautions must be taken against vandalism. In some circumstances an underground station may be the best solution.

Screens and similar equipment requiring regular maintenance are not suitable for small, isolated pumping stations. Rural stations, which must of necessity be left unattended for long periods, should be fully automatic and must have pumps capable of handling crude unscreened sewage.

The station must be designed so that additional units can be added to cater for future development if this is probable. If relevant, space should also be available for a further rising main at a later date. Any sewer which will take the discharge from the rising main must be checked to ensure that its capacity is adequate for the maximum pumping rate envisaged.

The modern sewage pumping station equipped with automatically controlled, electrically driven pumps is a very reliable installation and is now the most usual type of station for foul sewerage. Compressed air ejectors are frequently used for smaller flows.

Further information on pumping plant, buildings and rising mains is included in the author's *Pumping Stations for Water and Sewage* [64].

CHOICE OF PUMPS

Among the many varieties of pumps manufactured, those used for sewage are centrifugal, mixed-flow and axial-flow. For smaller flows, the compressed air ejector is very suitable, and for certain more specialized duties (particularly at sewage treatment works) use is made of reciprocating pumps, air lifts, and the more recently developed screw-type pumps.

Centrifugal pumps (radial flow pumps), are generally suitable for outputs of up to 1000 m^3/h for heads up to about 40 m. They are not self-priming, and as foot valves are only suitable for relatively clean liquids, if these pumps are to operate automatically they must be installed below the level of the sewage to be pumped. This normally entails the construction below ground of a 'dry well' to house the pumps, adjacent to the 'wet well' containing the sewage. Some submersible pumps are, however, available for use in the wet well itself, but in general pumps installed in a separate dry well are easier to maintain.

It is usually convenient to choose vertical spindle pumps and to install the motors at a higher level. In this way all electrical equipment can be kept above flood level. Pumps with built-in priming devices are available in the smaller ranges for automatic operation, thereby allowing the use of horizontal spindle pumps installed at ground level, but their use is limited. Fully submersible pumps are also available, where both pump and motor operate in the wet well.

Axial-flow pumps are more useful for lifting large quantities through a limited head (capacities over about 2500 m^3/h at heads of 6·0 to 18·0 m). They are unsuitable for use with liquids containing solids, and they therefore find more favour for pumping surface water and for recirculation of final effluent rather than for foul sewage duties. Axial-flow type pumps are used extensively for fen drainage work. They are of comparatively simple construction, and generally operate at higher speeds than centrifugal pumps.

Mixed-flow pumps are those in which the flow through the impeller is partly axial and partly radial. They are useful for storm sewage duties and are generally available for outputs up to 1400 or 1500 m^3/h, and for heads up to approximately 15 m.

Ejectors are normally self-contained, with their own automatically controlled compressors and 'air bottle'. Maintenance is fairly straightforward and, being very reliable, they are suitable for isolated houses or small groups of houses, and for low-level installations, such as basements and public conveniences.

The choice between a centrifugal, a mixed-flow, or an axial-flow pump is basically a matter of comparison of speed, output and head conditions. With the three types, the flow increases as the head decreases. The decrease in head is more rapid with axial-flow pumps than with centrifugal or mixed-flow pumps.

Pumps are conveniently compared by reference to their 'specific speeds'. Specific speed can be defined as that speed, in revolutions per minute, at which an impeller generally similar to the one under consideration, and reduced in size, will develop unit head at unit output.

$$n_q = \frac{n\sqrt{Q}}{H^{3/4}}$$ **Formula 14.1**

The terms in Formula 14.1 can be expressed in suitable units so that:

n_q is the specific speed in rev/min
n is the actual speed in rev/min
Q is the capacity of the pump in m^3/h
H is the total manometric head in metres

The three types of 'rotating element' pumps have specific speed ranges, in the above terms, as follows:

Centrifugal: up to 5000
Mixed-flow: 5000 to 10 000
Axial-flow: 10 000 to 20 000

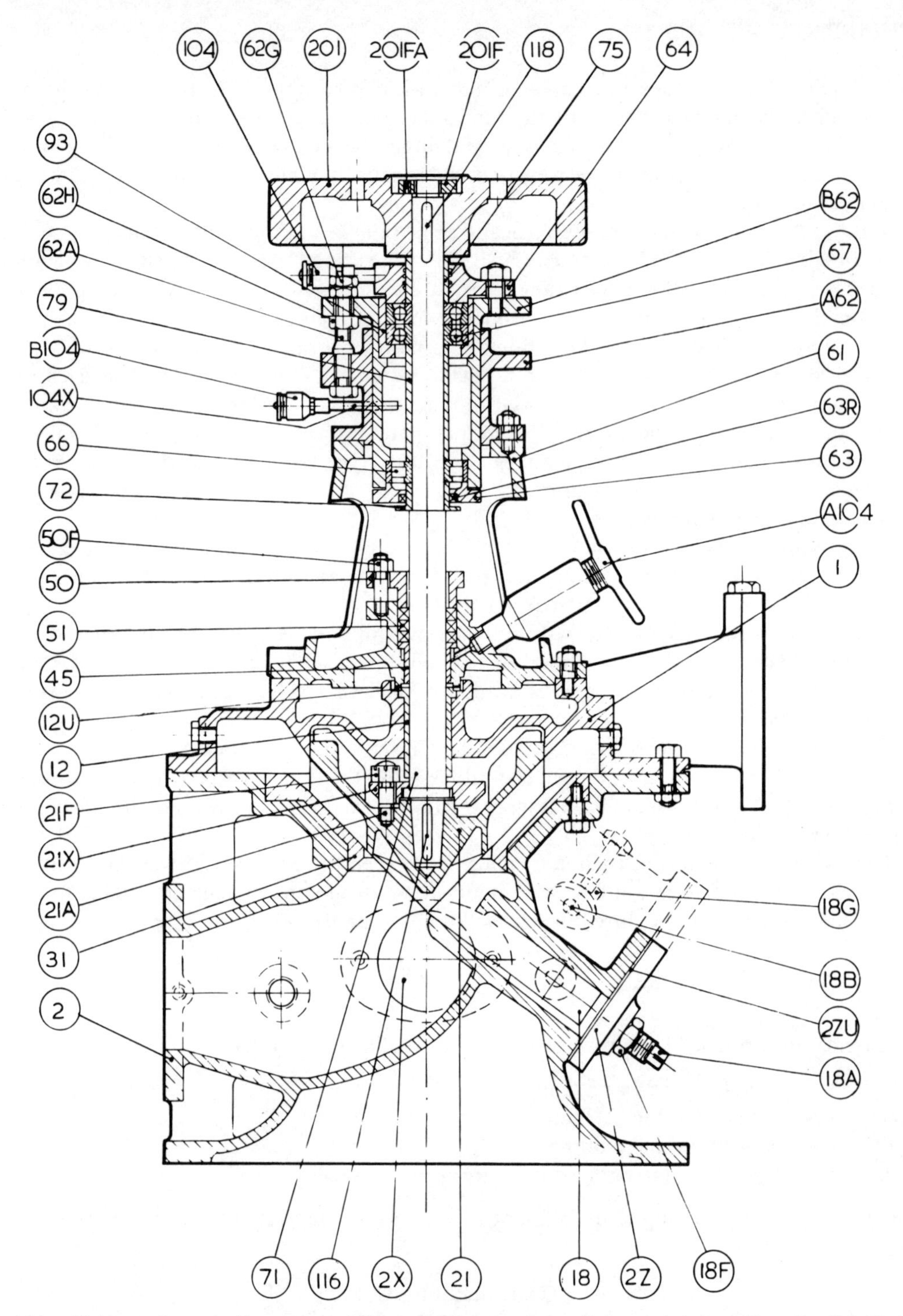

Fig. 14.1. *Disintegrating pump (by courtesy of Sigmund Pulsometer Pumps Ltd—see their Data Sheet No.* 43·7 *(V) for references).*

Although the specific speed is essentially a matter for the pump manufacturer, it is useful for the design engineer to have a general understanding of the subject, as it will give him an indication of the type of pump which will be required for a particular duty.

UNCHOKABLE PUMPS

When pumps are to be used for foul sewage, it is either necessary to ensure that they have special 'unchokable' impellers, capable of passing comparatively large solids, or the inlets to the pumps must be fitted with protective screens. Centrifugal pumps can be designed to be more or less unchokable, but the higher efficiency impellers of axial-flow and mixed-flow pumps must be protected if they are to be used for crude sewage.

No pump is entirely unchokable, but most sewage pump manufacturers produce a range of pumps in the smaller and medium sizes which will pass a 100-mm ball. This is generally considered the standard required for a pump to be suitable for crude sewage. The efficiency of these pumps is less than those with more complex impellers, but it is usually preferable to forgo some efficiency for the sake of reliability in service. For pump sizes up to about 250 mm (nominal suction diameter), the use of unchokable pumps is now almost universal when handling foul sewage.

The usual unchokable impeller has only one or two blades set tangentially across the eye so that there is no leading edge. It is the leading edge of a conventional impeller which tends to collect string and rags which can build up and eventually cause a blockage. In a truly unchokable pump the shaft will therefore not extend through the eye of the impeller, and it is supported by one radial and one radial-thrust bearing, the radial bearing being placed as close to the stuffing box as is feasible. When a pump is not of the 'split casing' type, a hand hole should be provided on either the pump casing or on the inlet bend to facilitate the clearing of any obstruction at the impeller.

BSCP 2005 recommends that 'for small to medium capacities, the centrifugal pump should be of the unchokable type, when any solid, up to a maximum of about 100-mm diameter sphere, that may enter the pump suction will be passed through the pump'. There is, however, a limit to the size of unchokable pumps, and for larger flows some form of screening will be necessary; these screens should preferably be of the automatically controlled, mechanically raked type, with bars at 'medium' spacing, i.e. about 10- to 20-mm clear opening.

DISINTEGRATING PUMPS

Various forms of self-clearing centrifugal pumps are available in which a cutting knife (or knives) works in conjunction with the impeller. The unit acts as both a disintegrator and a pump. The manufacturers claim that in addition to faecal matter, these pumps will deal with rags, pieces of wood and other hard materials. A typical cross-section of one version of this pump is shown at Fig. 14.1.

This type of pump is suitable for isolated pumping stations or where screening at the inlet would not be convenient. It is particularly suitable for coastal towns when crude sewage is to be discharged direct to the sea. Where these pumps are installed on a 'combined' sewerage system, the wear due to the grit in the sewage may be high and the extra cost of overheads and servicing should be considered.

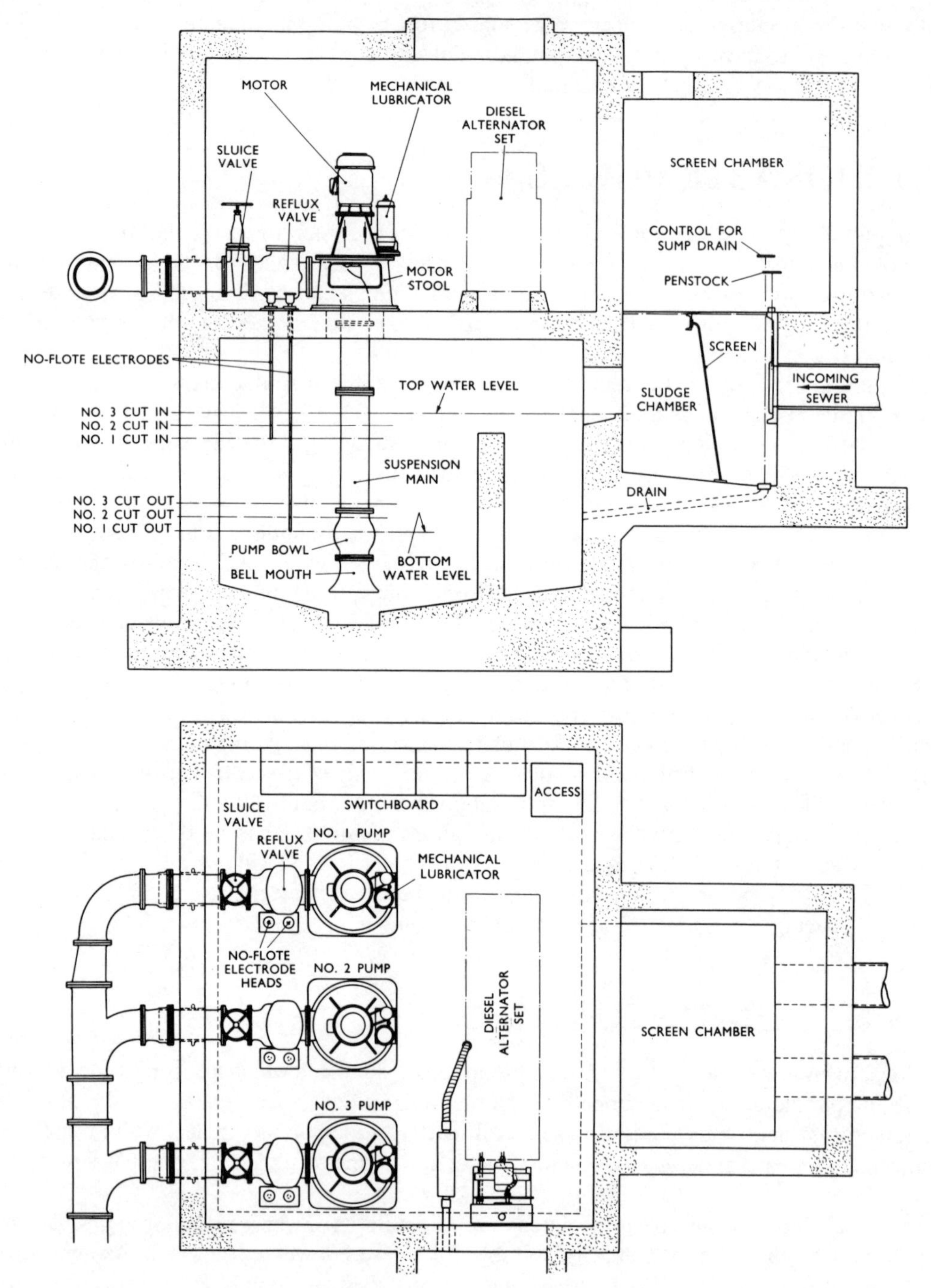

Fig. 14.2. A typical extended spindle pumping installation with suspended mixed-flow bowl pumps (by courtesy of A.P.E.—Allen Ltd).

The efficiency of the disintegrating pump is good within its recommended ranges of capacities up to 500 m^3/h or over, and at heads of up to 30 or 40 m.

EXTENDED SPINDLE PUMPS

These pumps are used to some extent for pumping sewage as the omission of a separate dry well reduces the size and cost of a pumping station, while at the same time the installation of the pumps below liquid level ensures that they are always fully primed for automatic action. These pumps are normally of the mixed-flow type and are suitable for storm sewage duties, when the pumps do not normally run for long periods and maintenance is therefore less frequent. Generally, they are more expensive than unchokable pumps installed in a conventional dry well, and maintenance is more difficult, as the unit must be completely withdrawn from the wet well before any work can be carried out. A typical layout of this type of pumping installation is illustrated in Fig. 14.2.

SUBMERSIBLE PUMPS

Completely submersible pump/motor units have now become very popular. There are two basic categories of pumps available, namely sewage and solids handling pumps or clean water pumps.

The sewage pumps are available either as transportable or permanently installed units and are capable of discharging up to 1400 litres/s and of achieving heads in excess of 50 m. A typical installation is shown in Fig. 14.3.

The clean water pumps are used extensively on construction sites for general dewatering purposes. The advantages of this type of pump unit have been recognized and they are now being

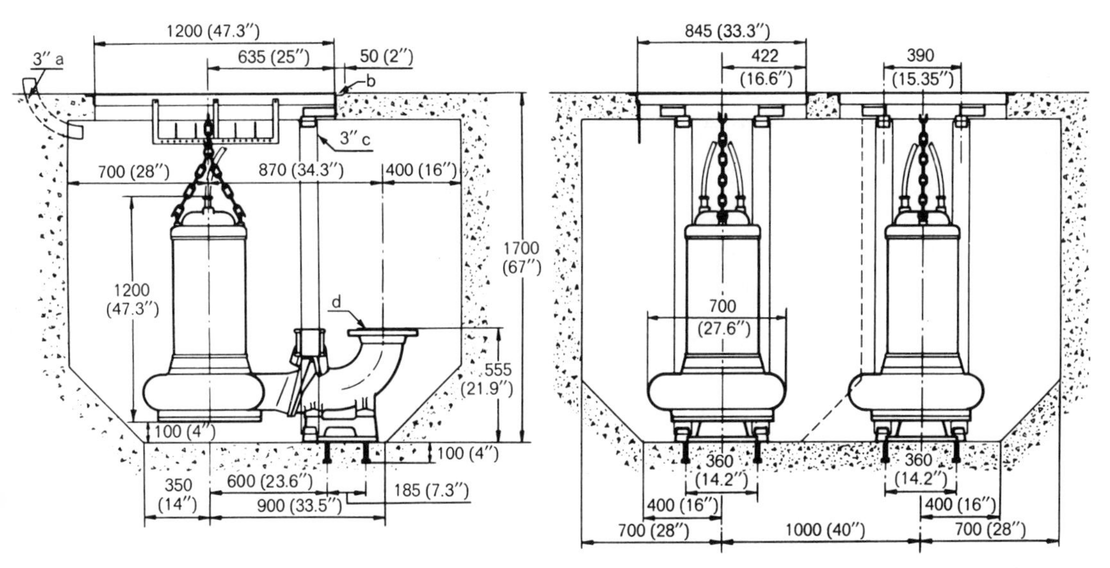

Fig. 14.3. Submersible pumps. Minimum dimensions for an installation of two pumps, capacity of each up to 600 m^3/h (by courtesy of Flygt Pumps Ltd).

used for specialist applications throughout industry, e.g. fish farming, oil rigs, food industry, vegetable processing, etc.

In addition to these two main groups, developments have been made into specialist areas and pumps are now available, manufactured in stainless steel, for chemical applications. Units are also available certified for use in flammable atmospheres, thus making them suitable for use in coal mines, etc.

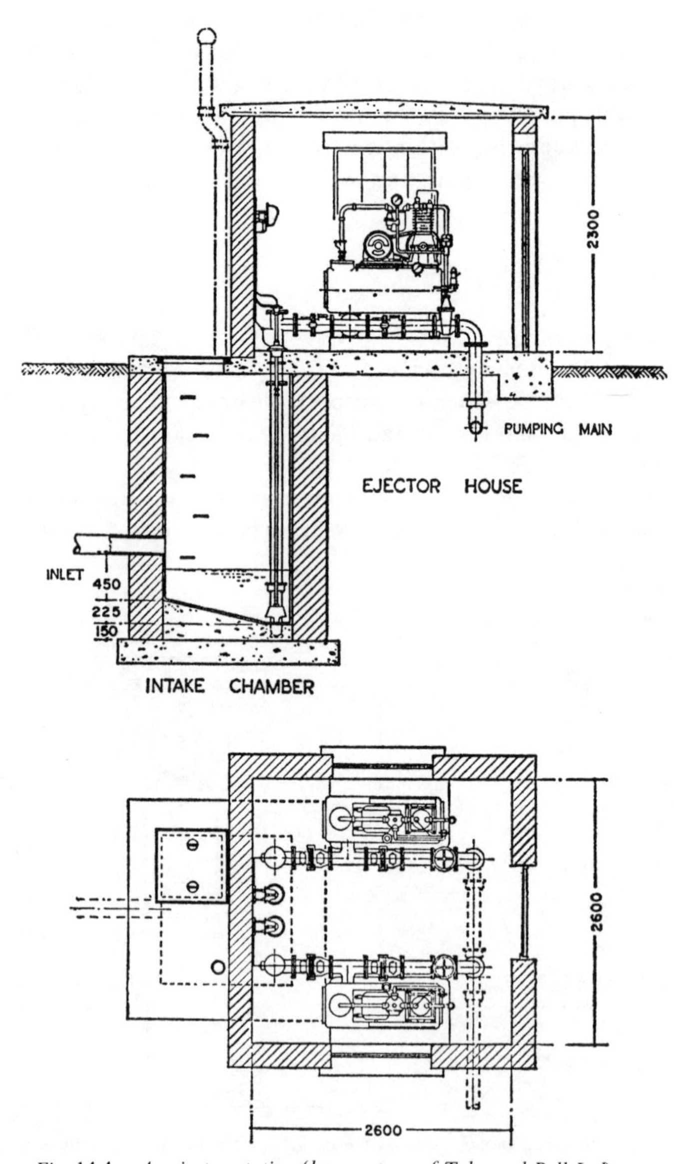

Fig. 14.4. *An ejector station (by courtesy of Tuke and Bell Ltd).*

EJECTORS

Compressed-air ejectors are available for flows between about 3 m^3/h (0·001 cumec) and 30 m^3/h (0·01 cumec) and can therefore be used to deal with isolated properties or with groups of up to 100 or more houses. Efficiencies may be as low as 20 % and are rarely over 50 %, but ejectors are very reliable, and as no sump or screens are needed they are often more suitable than small-capacity pumps for low flows. The actual rate of discharge is usually considerably greater than the nominal capacity of the ejector, and this should be taken into account when calculating the diameter of, and the velocity in, the rising main.

Compressed-air ejectors can be installed in a chamber built below the invert of the incoming sewer, and fed by gravity, or they can be of the 'lift-and-force' type, in which case the machinery is installed above ground level. An example of the latter type of installation is given in Fig. 14.4. Lift-and-force ejectors are more easily maintained, but there are situations in which a gravity-fed system may be preferable. Two small gravity-fed ejectors can be installed in an underground chamber about 2·5 by 2·0 m, while a larger pair would require about 3·0 by 2·5 m. A pair of small lift-and-force ejectors can be housed in a building about 3·0 by 3·0 m by about 2·3 m high.

Compressed air can be provided by a compressor at each ejector station, or it is possible to arrange for a number of stations to be fed from a central compressor station. In the low-lying, flat district of Rangoon, Burma, twenty-two ejectors were commissioned in about 1890, all served from one central compressor station. Installations of that type are not common, and it is now more normal for each ejector station to be self-contained. Ejectors are usually installed in pairs, and where a breakdown would have serious results duplicate air compressors should be provided.

SPECIAL INSTALLATIONS FOR SMALL FLOWS

In more recent years other types of installation have been developed for pumping small quantities of unscreened sewage without the use of compressed air. These include the Sigmund Pulsometer 'Solids Diverter' (see Fig. 14.5) and the Mono 'Mutrator'. They are suitable for flows from about 2 to 55 m^3/h. The 'Solids Diverter' consists of a mild-steel sewage receiver, twin electrically driven pumps (one duty and one standby) and a system of non-return valves. As only the liquid part of the sewage passes through the pumps (the solids are diverted) the efficiency is relatively high (up to 60 %); no further standby is necessary; and heads of up to 35 m are acceptable. The 'Mutrator' includes a macerator for the solids, is self-priming, can be installed at ground level with a suction lift of between 4·5 and 6·0 m and can be used with a small bore rising main (see Fig. 14.6).

Taking 100 mm as a minimum diameter for the rising main, and with a minimum velocity of 0·75 m/s, it will be apparent that the minimum pump output should be about 20 m^3/h.

SELF-CONTAINED 'PACKAGED' PUMPING STATIONS

A recent development intended for isolated housing estates and similar conditions, the self-contained packaged station houses the pumps and motors (with a wet well if required), together

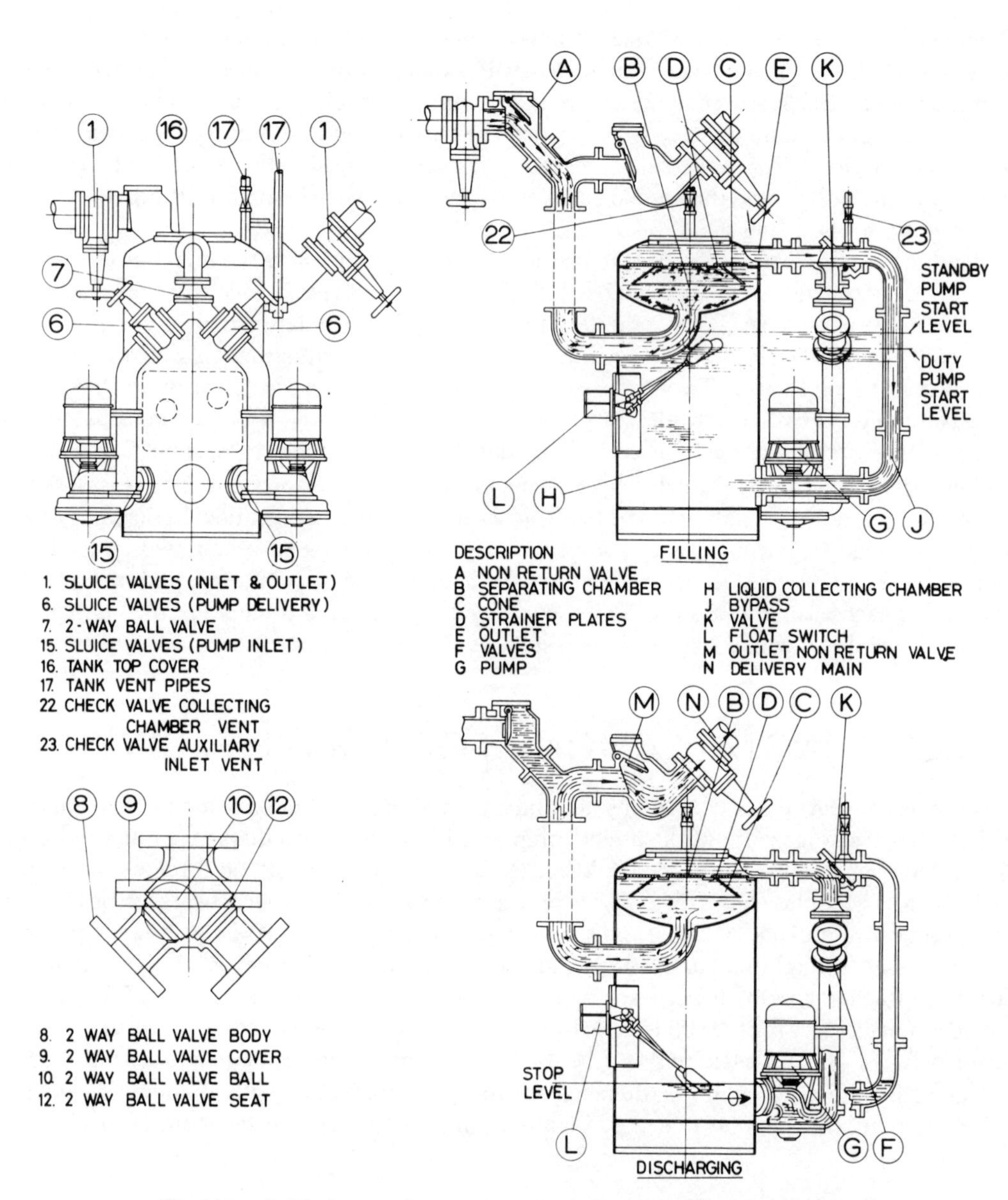

Fig. 14.5. Solids diverter (by courtesy of Sigmund Pulsometer Pumps Ltd).

Fig. 14.6. *The mono mutrator (by courtesy of Mono Pumps Engineering Ltd).*

with all valves and switchgear in one factory-built unit. This type of unit can now be supplied by a number of manufacturers, and is suitable for flows up to about 200 m^3/h. One type of packaged unit is illustrated in Fig. 14.7.

SCREW PUMPS

Screw pumps have been developed during recent years for lifting liquids through heads up to about 10 m (see Fig. 14.8). They are a modern development of the Archimedes screw, rotating slowly in an inclined trough. Screw pumps are available with capacities up to 25 000 m^3/h, and are claimed to have efficiencies of between 65 % and 70 %.

The design of this type of pump makes it completely unchokable, and the manufacturers claim that the screw action and the low speed (between 20 and 90 revolutions per minute), are ideal for lifting activated sludge, as the floc is not damaged. Screw pumps are available for a wide range of flows, and can be used for pumping both crude sewage and storm water.

A *Handbuch der Wasserforderschnecken*, by Messrs. Ritz-Pumpenfabrik OHG of West

Fig. 14.7. A packaged-type pumping unit (by courtesy of Pegson Ltd).

Fig. 14.8. *Screw pumps* (*by courtesy of Simon-Hartley Ltd*).

Germany (associated with New Haden Pumps Ltd) contains very full details on the design and installation of these pumps.

PUMP CAPACITIES

The pumps in a sewage pumping station must be suitable for a wide range of flows, from low (night) flows to peak daytime flows. If the sewage is stored in the pump well for long periods of time it will turn septic, resulting in the generation of hydrogen sulphide, with its objectionable smell, and making the sewage difficult to treat at the works. Except in small installations, it is therefore usual to install more than one duty pump, so that one pump will deal with low flows and further pumps will come into operation as the flow increases. If the rising main will discharge at or near the treatment works, special consideration must be given to the effect of pumped flows on the operation of the works. Pump capacities should be chosen to even out the surges as far as possible. The use of variable-speed motors will overcome surge problems to a great extent, but they are, of course, more expensive and more complicated to maintain than single-speed motors.

Some designers prefer to have several sizes of pump to cater for variations in flow, with each pump cutting out in succession as the next larger pump starts up. This is not generally considered good practice, and, as each motor must be electrically interconnected, a fault in the relay system could result in the whole station being out of action. It is better to have only one (or at the most two) sizes of pump in a station. In this way, the 'duty pump' can be changed regularly to avoid excessive wear on one pump, spares are more easily available, and the wet well and rising main can be designed more accurately.

To be truly interchangeable, the pumps and motors should be similar in all details, including the 'hand' of the pump. A pump is 'left-hand' when the direction of rotation is anti-clockwise when viewed from the inlet branch (i.e. when viewed from *below* if a vertical-spindle pump). Right-hand pumps are available, but are not so common.

Standby pumping capacity must always be provided. A small station will have one duty pump and one standby, while larger stations must have sufficient installed capacity to cope with the flow if one of the larger pumps is out of action.

CALCULATIONS

Although the sewers of a 'separate' system may occasionally be designed for a maximum flow of 4 d.w.f., it is usual to install pumping capacity for up to about 6 d.w.f. If it is intended to install two sizes of pump, and the d.w.f. is, say, 0·125 cumec (450 m^3/h), the pump duties could then be divided as follows:

One pump type 'A'	500 m^3/h	1·1 d.w.f.
Two pumps together (type 'A')	950 m^3/h	2·1 d.w.f.
Three pumps together (type 'A')	1 350 m^3/h	3·0 d.w.f.
One pump type 'B'	1 600 m^3/h	3·6 d.w.f.
Two pumps together (type 'B')	2 800 m^3/h	6·2 d.w.f.

It is well known that the maximum load on the motor starters (in terms of the number of starts per hour) occurs when the inflow to the station is 50 % of the outflow. It can be shown that under those conditions:

$$Q = \frac{D \times t}{240}$$ **Formula 14.2**

where

Q is the storage capacity in m^3
D is the discharge in m^3/h
t is the total time in minutes to empty and refill the sump

If the starter capacity is limited to 15 starts per hour (BSS 587, 'Intermittent Duty'), then t must be not less than 4 min, so that:

$$Q = \frac{D}{60}$$ **Formula 14.3**

As D is the discharge per hour, the suction well must then have a capacity of at least 1 min discharge of the pump, the capacity being measured between the cut-in and cut-out levels of each individual pump. The length of the sump will normally be fixed by the plant layout, and as the depth of liquid between cut-in and cut-out levels of each pump should normally be between 450 and 600 mm, it is then comparatively simple to calculate the required width of the wet well. While the design of the pumping station is usually based on the use of 'intermittent duty' starters, many engineers allow a further factor of safety by specifying 'frequent duty' starters, suitable for up to 40 starts per hour.

Having established suitable pump outputs for the particular installation, the rising main should then be designed so that pumps of high efficiency can be chosen for the total head (static lift plus friction head), and so that the velocity in the main and the total head on the pumps are kept within accepted limits (see Chapter 15). The maximum possible static lift will be the suction lift from the lowest water level, plus the delivery head to the highest point of discharge. It may, however, be preferable to design the pumps so that their maximum efficiency occurs at a head less than this. The suction lift itself should be considered to avoid cavitation caused by the vaporizing of the liquid under partial vacuum when pumping against an excessive suction head.

When the output and total manometric head have been calculated, the pump power expressed in kilowatts can be found from the following formula:

$$P = \frac{Q + H}{3{\cdot}67\,r}\,\text{kW}$$ **Formula 14.4**

where

Q is the output in m^3/h
H is the total manometric head in metres
r is the efficiency expressed as percentage

If the total head on one of the type 'A' (500 m^3/h) pumps referred to above is 21·4 m and the pump efficiency is 40 %, then the power of the required pump is:

$$P = \frac{500 \times 21{\cdot}4}{3{\cdot}67 \times 40}\,\text{kW}$$

$$= 73\,\text{kW}$$

To calculate the required size of motor, further allowance must then be made for the efficiency of the motor. If this is, say, 85 % in the above example, then the motor power must be:

$$73 \times \frac{100}{85} = 86\,\text{kW}$$

MOTORS

While pumps may be driven by petrol motors, or oil and gas engines, the most usual prime mover for a sewage pump is the electric motor. Those used to drive vertical spindle pumps in the conventional station with wet and dry wells are normally of the 'drip-proof' type, although in some installations it may be preferable to use the more expensive 'totally enclosed' motors.

Most sewage pumps are driven by fixed-speed motors, although variable-speed motors may be used for special conditions. When the electricity supply is alternating current, motors may be either of the squirrel-cage or the slip-ring type.

Squirrel-cage motors are cheaper and are usually satisfactory in small installations and when a high starting current is acceptable. For direct-on-line starting, this current may be up to six times the full load current; this will be less with star-delta or auto-transformer starters.

Slip-ring motors, with stator-rotor starters, take about 1·25 times full load current for starting. Most electricity charges are now based on a kW charge for current used, plus a kVA charge which takes into account the magnitude of the starting load, and for anything except very small installations the use of slip-ring motors and stator-rotor starters is fairly common.

In installations where a variable output is required, it is possible to install pumps with variable-pitch impellers, or the motors can be of the variable-speed type. The latter can be achieved with either a slip-ring motor with rotor-resistance control, or by using a commutator machine with an induction regulator.

The 'power factor' of the motors will depend on their power and the load, and may vary from about 65% to over 90%. These figures can be improved by using power-factor correction condensers. The capital cost of these condensers can often be recovered very quickly by the saving in kVA charges.

CONTROLS

Recommendations for standards of control gear are set out in BSCP 2005. As the majority of sewage pumping stations incorporate electric motors using 50 Hz, 3-phase, 415-volt, A.C. current, this is the only type of installation considered here.

It is usual to arrange for the automatic starting and stopping of pumps, either by floats or with floatless control gear (electrode or pneumatic), operated by the level of the sewage in the wet well, so that the higher the level of sewage in the well, the greater the pumping capacity in operation. To allow the 'duty' pump to be changed at regular intervals, it is normal to include either a plug board or a hand-operated sequence changeover switch. The latter has a neater appearance. Where a chart-type level recorder is installed, it is now possible to arrange for the cutting in and out of pumps by the addition of microswitches, tripped by the stylus arm of the indicator. Figure 14.9 shows an installation of this type of control.

The use of floats for pump control is now quite limited, and pneumatic devices are not common in the UK. Electrode rods are the most usual form of control. These rely on the conductivity of the sewage, and as each electrode is covered or uncovered a relay is operated, which in turn controls the pump starter. Difficulties can arise with the build-up of grease and rags, so that either the relays fail to operate or they can be short-circuited. Special 'low sensitivity' controllers are available which, the manufacturers claim, overcome problems caused by electrode fouling.

One firm of switchgear manufacturers has produced a patented control system using only one pair of electrodes, in combination with motor-driven controllers with associated 'increase pumping' and 'decrease pumping' timers and relays. The manufacturers claim that with this method of control there is a reduction in depth of the wet well, with a corresponding reduction in construction costs.

Both float-type and floatless control gear should be carefully sited in the wet well to avoid the

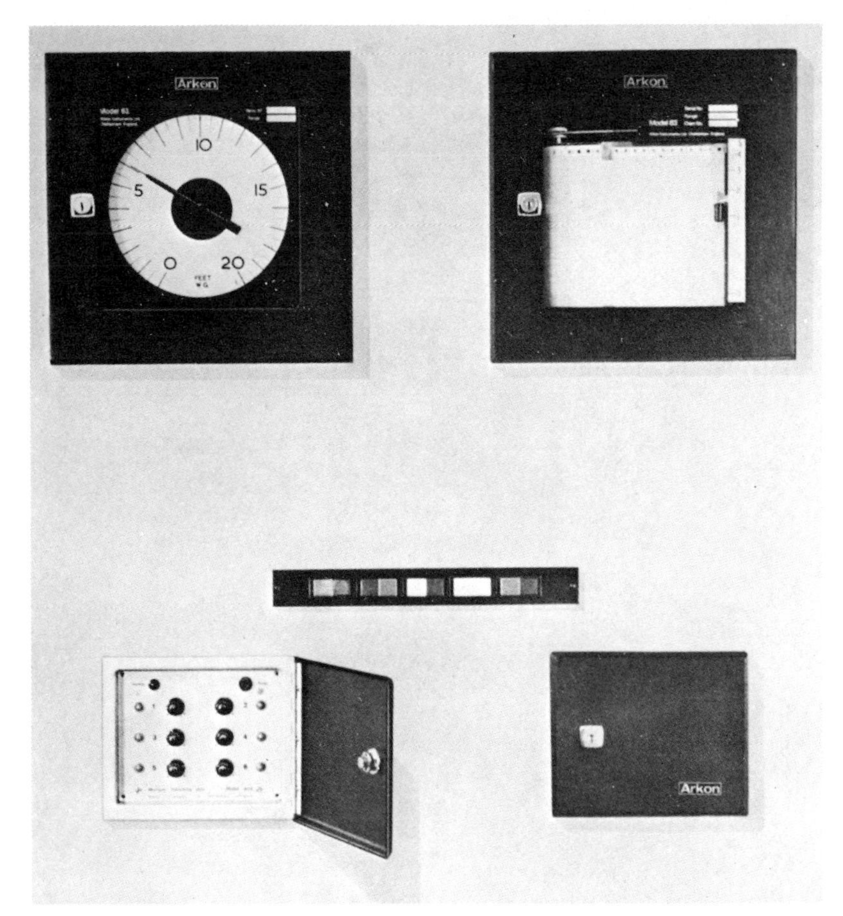

Fig. 14.9. *Pump control by micro-switches. A panel incorporating indicator, chart recorder and switching unit (by courtesy of Arkon Instruments Ltd).*

effects of vortex action when a pump is operating. Starting levels between individual pumps should be at not less than 150-mm intervals to allow the first motor to develop its full speed and output before the second motor starts. This is particularly important when large pumps are installed. Provision must always be made for manual starting as an alternative to automatic starting; facilities for manual starting are necessary during the commissioning of the installation.

The size of the panel housing the switchgear and controls will often influence the layout of the pumping station, as cable runs will be required, and openings in the walls and the access to the dry well must all be co-ordinated with the layout of the plant. Early discussion with manufacturers of switchgear and controls is essential, and in some circumstances it may be advisable to obtain tenders for these panels in the early stages of design, and ahead of tenders for the remainder of the equipment. This will then allow a more realistic approach to the design of the machinery layout, and of the building itself.

The amount of switchgear required at a pumping station will vary from simple wall-mounting starter panels (one per motor) for a small installation to composite floor-mounting units with

Fig. 14.10. *A composite panel of control gear (by courtesy of Whippendell Electrical Manufacturing Co. (Watford) Ltd).*

sections for main isolating switches, electricity supply meters, single-phase supply equipment for station heating and lighting, relay equipment for the level controls, and hydraulic meters and recorders, in addition to cubicles for the pump starters. Figure 14.10 shows a typical large installation.

Wherever possible a station should have two sources of power supply, so that a failure of one supply will not put the station out of action. Alternatively, an automatic diesel engine drive can be arranged for at least one of the pumps; that pump is then normally driven by an electric motor, but the diesel drive automatically starts up and takes over in the event of a power failure. As the diesel engine has to be started by compressed air, this type of stand-by must be very carefully designed to ensure that the engine runs for a sufficient time to replenish the air 'bottle'. Alternatively, separate arrangements must be made for this. For important installations, an emergency lighting system (operated by batteries) may be worthwhile; the batteries are kept charged from the mains supply and the lighting only operates when the main power supply is off.

As an alternative to, or in addition to, the installation of a flow recorder, each motor can be fitted with simple 'hours run' and 'operations' counters. These counters do not give any indication of the actual pattern of pumping, but are very useful guides for maintenance.

PUMPING STATION EQUIPMENT

The penstocks controlling the flow of sewage to the compartments of the wet well can either have 'rising' or 'non-rising' spindles, depending on the method chosen for their operation. Spindles should be extended to ground level so that they can be operated either from inside the motor room or from the roof slab of the wet well. If headstocks are fitted, rising spindles may be preferable, but if the penstocks are to be key-operated, non-rising spindles are more usual. If headstocks are fitted outside the building these should be locked to prevent unauthorized operation.

Within the station, pipework should normally be of flanged pipes, finishing at a flanged socket just outside the building. This is then a convenient terminal point for the pump manufacturer's contract, as the pipeline beyond this point will generally be of standard spigot-and-socket pipes. Flanged pipes are provided inside the building to prevent movement and leakage, and these should either be supported on concrete supports or by wall brackets. A safe span for cast-iron pipes can be calculated from the formula:

$$L = 484\,D^{0.5}$$ **Formula 14.5**

where

L is the span in millimetres
D is the diameter in millimetres

In practice, however, more supports are generally provided than are theoretically required. Suggested maximum spacings for supports for both horizontal and vertical runs of pipes of various materials are given in Table No. 8 of CP 304, 'Sanitary Pipework above Ground'. That table quotes a maximum spacing for supports for horizontal cast-iron pipes as 1800 mm for any diameter.

Reflux valves should be fixed on *horizontal* sections of pipeline and should preferably be fitted with external levers for opening manually to clear solids; these also enable the pumps and suction pipes to be back-flushed. Sluice valves should be suitable for use with sewage, with the provision of

Fig. 14.11. A typical dry well pumping station layout (by courtesy of A.P.E.—Allen Ltd).

either an access door or a scour plug for cleaning, and should not be buried underground unless this is unavoidable. When installed inside a building, valves should have external screws and should be operated by handwheels. Wherever possible, spindles should be brought up to motor-floor level and fitted with headstocks. A typical dry well installation is shown in Fig. 14.11.

As most pump glands leak to some degree, it is usual to drain the dry well floor towards a sump (floor slope about 1 in 100 minimum), which can be emptied either by a specially installed sump pump or by an auxiliary suction. The latter is a small suction pipe of about 30- or 40-mm diameter, connected into the main suction of a pump (upstream of the sluice valve on the main suction pipe) and operated by a full-way (plug cock) valve. If a sump pump is installed it should operate automatically, controlled by the level of the water in the sump. The floor can be kept dry by fitting small drain pipes from the glands of each pump to discharge into the sump.

Where vertical spindle pumps are installed with motors in a motor room more than 2 or 3 m above the pumps, it may be necessary for the pump manufacturer to incorporate intermediate shaft bearings. It is usual for rolled steel channels to be built into the station, spanning the pump chamber, to which these bearings are secured. Normally one channel is fixed for each intermediate bearing. For small or medium-sized stations, these can usually be 152 × 76 mm × 17·86 kg/m channels.

Pumps should be fitted with air-release cocks to operate automatically to release, from the pump casing, air which would otherwise be trapped while the pump is starting. They should be fitted with extension pipes so that they discharge into the wet well or outside the station.

A pressure gauge should be connected to each pump outlet, the gauge dials themselves (suitably corrected for height), being mounted in the motor room. A vacuum gauge should be fitted to each suction pipe as near the level of the pipe as possible. Connecting pipes to the gauges should be of heavy-gauge copper or nylon tube, and complete with two-way cocks, so that they can be cleaned of any obstruction without the need for dismantling.

Except for the smallest stations, overhead lifting gear should be provided for the installation or removal of plant, and, depending on the weight of the units to be handled, this may take the form of a travelling gantry crane or a travelling hoist on a fixed runway beam. At very large stations it may be desirable to include, within the superstructure, a loading bay of a size to take a fairly heavy lorry (CP 2005).

It is general to arrange for the motors to be bolted to rolled-steel joists fixed in the motor room floor, and for openings to be left for the easy installation and removal of pumps. If a simple fixed runway beam is provided, this should then be directly over the centre line of the pumps and motors.

At the time of commissioning a pumping station provision should be made for the supply of a set of tools for the pumps and motors, and also a reasonable supply of spare parts. Spares to be stocked will depend on the locations of the pumping station; an installation overseas or in a remote part of the country should have a wider range of spares than installations nearer the manufacturer's works. The minimum spares should include starter panel contacts, fuses, pump packings, and probably pump bushes and glands. The inclusion of a strongly bound log book at this stage should ensure that a log is maintained from the initial commissioning.

Open-type flooring (see Fig. 14.12) can be used for stairways and intermediate platforms, but is not suitable for use at motor-floor level, as a spanner or bolt dropped through the flooring can injure a workman or can damage machinery at a lower level. Removable sections of flooring should preferably be of steel or aluminium chequer plating, cut to sizes and shapes that can easily be lifted, and set into angle frames which provide a stop for motor floor finishes (quarry tiles, etc.). A table of weights and safe loads for chequer plating is included in Chapter 4.

Hand-railing should be fixed around all openings (including the stairwell), with gates or removable sections of chains where required. This hand-railing can be either of heavy-duty steel or aluminium and is generally of the two-rail type with the upper rail about 1 m above floor level.

In an isolated pumping station a telephone is desirable, as it will enable the maintenance fitter to keep in contact with his office, and, in the event of injury or other emergency, will ensure that assistance can be made available quickly. A first-aid kit should also be provided.

BUILDINGS

Apart from the obvious road reinstatement, pumping station buildings are probably the only parts of a sewerage scheme which are left visible to the general public after the completion of a contract. They represent a comparatively small part of the total cost of the scheme, but often the designer's ability will be judged by their efficiency and appearance.

The site of a station will be determined by the overall sewerage scheme, but it will, however, be apparent that minor adjustments in siting will be possible without any material effect on the design. Careful consideration should be given to the final location to ensure that the building blends into

Fig. 14.12. Open-type steel flooring (by courtesy of Steelway (Glynwed Integrated Services Ltd)).

the district. An adjustment of, say, 50 m may make a considerable difference to the amenity of the locality. The site should, of course, be free from liability to flooding and should be readily accessible for construction and maintenance.

The dimensions of the building will be set by the type of plant to be installed, and the minimum height will be dictated by the requirements of any lifting gear. Despite this the final dimensions should match the general pattern of the locality and the style and finish of the building should be chosen depending on whether it will be in the country or in a built-up area. Money spent on the careful choice of facing bricks or roofing tiles is usually well spent.

A pumping station will generally consist of two parts, the substructure and the superstructure. While the substructure is basically a matter for engineering design, the superstructure should preferably be designed by an architect. If the substructure is divided into wet and dry wells, it is usual for the dry well to be directly under the superstructure and to have access from inside the building. Access to the wet well should be from the open air. The structural design of pumping stations is referred to in BSCP 2005.

The substructure will normally be constructed of either mass or reinforced concrete. BS 5337, 'The Structural Use of Concrete for Retaining Aqueous Liquids', sets out standards for a reinforced concrete structure. As with the design of manholes, it is usually necessary to check that the structure will not float when empty. If this is possible this may entail thickening up the floor slab or walls, or extending the floor slab beyond the walls to utilize the weight of the surrounding earth.

Pumping Stations for Water and Sewage gives general guidance for determining the possible size of pumping units and pipework, so that a preliminary layout can be prepared.

The wet well should preferably be divided into two compartments either of which can be isolated from the incoming flow for cleaning. The incoming sewer should then discharge through an inlet chamber to either or both compartments. The floor of the wet well should slope at 1 in 1, or steeper, to a sump at the pump suctions. Many engineers prefer a slope of about 1·75 vertical to 1 horizontal, but on the other hand these steeper slopes are inconvenient for maintenance. It may be preferable to construct a long, deep channel at the suction pipes, with a benching sloping at about 1 in 6; if this form of construction is used the lowest pump cut-off level should then be in the channel itself (i.e. below the level of the benching).

The suctions should preferably be not more than $2D$ apart (where D is the diameter of the suction) and, to avoid the deposition of solids between the suctions when turned down, the distance of the suction above the floor should be between $D/2$ and $D/3$.

The recommended capacity between cut-in and cut-out levels of pumps has been referred to earlier. The retention time in the wet well should be kept as short as possible and should preferably not be more than 30 min. The lowest water level should be at least 600 mm above the lip of the bellmouth to ensure that the pumps are primed at all times. Access to the wet well should be large enough to facilitate cleaning down inside, and adequate ventilation must be provided, preferably with special ventilators.

The superstructure will normally be of brickwork, with either a pitched tiled roof or a flat roof. The roof must be high enough to accommodate any lifting beam together with the lifting tackle. Sufficient space must be allowed for the vertical lift required; this may be quite considerable with extended spindle pumps. For very small stations, factory-made 'cubicles' are available, while for some installations no superstructure is required beyond perhaps a small cabinet to house the electrical switchgear.

Recommendations for the co-ordination of dimensions in building were published in BS 4011,

1966, 'Basic Sizes for Building Components and Assemblies'. The four preferences for co-ordinating dimensions were then given as:

i. n × 300 mm
ii. n × 100 mm
iii. n × 50 mm
iv. n × 25 mm

These co-ordinating dimensions are being used as far as practicable by BSI in the preparation of new standards and by suppliers of building materials. The Ministry of Public Building and Works has issued recommendations for both horizontal and vertical dimensions for various types of buildings; these adopt the 300 mm module for longitudinal dimensions and the 100 mm and 50 mm modules for vertical dimensions.

BS 4011 states:

> The basic sizes for the co-ordinating dimensions for components and assemblies should be chosen after consideration of the relevant functional requirements. Within each category, such as windows, door frames, wall panels, floor slabs, etc., the preferred sizes should as far as possible be the same in all materials. Account should be taken of the need for different categories or components or assemblies to occupy building spaces of the same size. Where a number of components are used to build up an assembly, the overall size for the assembly should be a basic size.

Unfortunately, it is usually necessary to take precautions against vandalism at isolated buildings. Windows can be set high in the walls, while doors and locks should be more solid than required for normal industrial or domestic usage. The use of toughened glass (minimum recommended thickness, 5 mm) will reduce the possibility of damage to windows. Special glass or plastic-filled reinforced-concrete blocks may be suitable for use either in the walls of the superstructure or as 'pavement lights' over the wet well (see Fig. 14.13). Unreinforced glass blocks are non-load-bearing, but will carry their own weight and can be used, provided suitable framing supports are incorporated.

Fencing should always be provided around a pumping station to protect the equipment. Except at very small stations, a pair of double gates and an access road (minimum, 3·5 m wide) should be provided to facilitate the installation and maintenance of the machinery. The site within the fence should be landscaped as necessary to blend with the surrounding area. This should generally be as simple as possible to reduce maintenance costs.

Washing and toilet facilities should be provided at all stations which are to be regularly attended. A small electrically operated water heater should be included. Drainage of these facilities can, of course, be taken to the wet well in a foul sewage pumping station. Roof and access road drainage can often also be taken to the wet well, or it may be preferable to discharge this to a neighbouring ditch or stream. The provision of a tap and hosepipe will ensure that the well can be hosed down regularly; this will not only improve working conditions in the well, but will also remove accumulations of solids which might cause septicity and subsequent odours.

Heating is necessary in the motor room to prevent condensation in electrical motors and switchgear. This may take the form of special anti-condensation heaters in the equipment itself, but in addition it is usual to provide wall heaters, as these will prevent the freezing of the plumbing in cold weather and also maintain a comfortable temperature for any attendants. These heaters should be thermostatically controlled.

Fig. 14.13. *Installation of pavement lights (by courtesy of J. A. King & Co. Ltd).*

Lighting in the dry well and in the superstructure must be adequate for proper operation and maintenance of the equipment. As stated in CP 2005, 'the siting and intensity of the permanent lighting installation should be such that the use of land lamps is seldom necessary, other than for close inspection'. Any permanent lighting in the wet well should be controlled from switches in the motor room.

All openings in floors of pumping stations should be provided with either hand-railing or covers. Rotating shafts, chain drives, etc., must be protected with guards; rubber mats should be used in front of electrical switchgear panels, especially if these are not of the enclosed type. Stairways should always be incorporated in preference to ladders (except in very small stations) and stair design should follow as far as possible the accepted maxim that twice the riser plus the tread ($2h + d$) should be between 600 and 630 mm, where d is 250 mm minimum and h is 190 mm maximum.

15 Rising Mains

RISING mains must be designed to withstand the total manometric head on the pumps (static head plus friction losses), as this is the head which will apply in the vicinity of the pumping station. An additional allowance should be made for the effects of surge in cases of high pumping velocity, or in very long rising mains. Should the pipeline cross a valley at a lower level than the pumping station itself, the static head at the crossing may be greater, and this should be borne in mind when assessing the total head on the pipeline.

Materials used for rising mains include iron, steel, asbestos-cement and plastics. Vitrified clay and concrete pipes (except prestressed concrete pipes) are not suitable. Ductile iron (spheroidal graphite iron), steel and plastic pipelines are particularly suitable where ground movement is possible, but, quoting BSCP 2005, 'in the case of steel pipes, increased wall thickness and protective lining may be necessary to allow for possible corrosion, and the pipes should be wrapped with a protective sheathing'.

HYDRAULIC DESIGN

While tables such as Crimp and Bruges [68] and Escritt [70] may sometimes be used when designing rising mains, it is normal to restrict their use in this respect to any early approximations. Pumping plant design must be based on an accurate assessment of the total manometric head, and it is therefore usually necessary to calculate the friction head in rising mains as accurately as possible. It may sometimes be convenient to calculate the friction head in the main for new pipes, and again for the pipes after they have been in use for a few years.

Most pump manufacturers base the performance of their pumps on the friction losses in new pipes. Various manufacturers use different formulae, but the losses are normally calculated on the basis of 'smooth and new' cast-iron or steel pipes and 'clean' asbestos-cement or PVC pipes. While the friction losses in new pipes must be known if the pump outputs are to be tested on commissioning, the designer is really more interested in the working conditions later, when the pipes are 'old'.

Of the numerous formulae which have been devised, the one by Williams and Hazen gives a good degree of accuracy, and is simple to use. In metric terms, the Hazen-Williams formula for friction head is:

$$h = \frac{1128 \times 10^9}{d^{4 \cdot 87}} \times \left\{\frac{Q}{C}\right\}^{1 \cdot 85}$$ **Formula 15.1**

where

h is the friction head in metres per 1000 m
Q is the flow in m^3/h
d is the inside diameter of the main in millimetres
C is a friction coefficient (see Table 15.1)

TABLE 15.1
FRICTION COEFFICIENTS FOR USE WITH THE HAZEN-WILLIAMS FORMULA

Pipe material	*Condition*	*Value of 'C'*	
		300-mm diameter and under	*Over 300-mm diameter*
Uncoated cast iron	18 years old	100	—
Uncoated cast iron	10 years old	110	—
Uncoated cast iron	5 years old	120	—
Uncoated cast iron	Smooth and new	125	130
Coated cast iron	Smooth and new	135	140
Coated steel	Smooth and new	135	—
Uncoated steel	Smooth and new	140	145
Coated asbestos-cement	Clean	145	150
Uncoated asbestos-cement	Clean	140	145
PVC	Clean	150	—

The velocity of flow according to the Hazen-Williams formula is:

$$v = 10{\cdot}93 \times 10^{-3}\, C\left(\frac{d}{4}\right)^{0{\cdot}63} . s^{0{\cdot}54} \qquad \textbf{Formula 15.2}$$

where
- v is the velocity in m/s
- s is the hydraulic gradient

The Manning formula in its *original imperial* form was:

$$v = \frac{1{\cdot}4858}{n} m^{2/3} i^{1/3}$$

This can be expressed, to give the friction head, as follows:

$$h = 7952 \times 10^{9} \left\{\frac{Qn}{d^{8/3}}\right\}^{2} \qquad \textbf{Formula 15.3}$$

where
- h is the friction head in metres per 1000 m
- Q is the flow in m^3/h
- d is the inside diameter of the main in millimetres
- n is a friction coefficient (see Table 15.2)

A formula for *new* bitumen-lined steel pipes quoted by Messrs. Stewarts & Lloyds for clean water is:

$$h = 136{\cdot}9 \times 10^{6} \times \frac{Q^{1{\cdot}8}}{d^{4{\cdot}85}} \qquad \textbf{Formula 15.4}$$

TABLE 15.2
FRICTION COEFFICIENTS FOR USE WITH MANNING'S FORMULA

Pipe material	*Condition*	*Value of n*
Coated iron and steel pipes	New	0·010
Uncoated iron pipes	New	0·011
Galvanized iron pipes	New	0·012
Vitrified clay pipes	After some weeks of service	0·012
Iron and steel pipes	After use	0·013
All pipes	With imperfect joints and in bad condition	0·015

where
h is the friction head in metres per 1000 m
Q is the flow in m^3/h
d is the inside diameter of the main in millimetres

Stewarts & Lloyds point out that the effect of slime growth inside pipelines may increase the friction head by up to 50 % over that calculated by the formula. An additional allowance must also be included for friction in fittings.

TAC Construction Materials Ltd recommend the use of the formula developed for use with asbestos-cement pipes by Professor Scimemi. The formula for velocity of flow is:

$$v = 165\, m^{0 \cdot 68} i^{0 \cdot 56}$$ **Formula 15.5**

where
v is the velocity of flow in m/s
m is the hydraulic mean depth in metres
i is the hydraulic gradient

Formula 15.5 can be transposed to give the friction head in *new* asbestos-cement pipes as follows:

$$h = 91 \cdot 87 \times 10^6 \times \frac{Q^{1 \cdot 785}}{d^{4 \cdot 78}}$$ **Formula 15.6**

where
h is the friction head in metres per 1000 m
Q is the flow in m^3/h
d is the inside diameter of the main in millimetres

It is recommended that 5 % to $7\frac{1}{2}$ % should be added to the friction losses calculated by Formula 15.6 to allow for joints and imperfections in laying the pipes.

VALVES AND FITTINGS

A rising main is a pressure pipeline, and its design will differ in many respects from the design of a gravity sewer. The basic principles of structural design will equally apply, and to some extent the hydraulic design is similar, but in addition, as the pipeline will normally be laid to follow the ground contours (minimum cover generally about 1·0 m), provision must be made for the release of trapped air at high points, and for washing out at low points. Air valves suitable for use with sewage must be provided at all high points on the line, and may also be required at intermediate positions along long lengths of even gradient.

A washout should take the form of a tee-junction and valve, and should discharge into a gravity sewer if possible. Valves and connecting pipelines should be at least 80-mm diameter. Where no suitable sewer is available, the washout may have to discharge to a specially constructed sump, which must then be emptied after use. The section of the rising main near the pumping station can sometimes be arranged so that a washout can discharge to the wet well of the pumping station; if so, the wet well must be of adequate capacity, or alternatively, if a second rising main is installed, the pumps can be used to control the level in the well during the period of washing out.

Hatchboxes have been installed on rising mains in the past to provide access points at bends, valves, etc., but these are now rarely used. Some engineers install short lengths of pipe with victaulic or similar joints at 300- or 400-m intervals in lieu of hatchboxes. These can be built into special brick chambers or they can be covered over and carefully referenced.

Sluice valves and reflux valves are fitted to each pump outlet, and further valves are often not necessary unless it is intended to be able to isolate the rising main for washing out purposes or for the addition of further pumping units in the station at some future date. If so, it is recommended that a sluice valve and reflux valve be installed in a chamber immediately outside the pumping station, in addition to the valves on the pumping units themselves.

If twin rising mains are installed, an emergency by-pass should be provided between the two pipelines, and sufficient sluice valves must then be fitted immediately outside the station, so that either main can operate as the duty main or both pipelines can operate either separately or together.

The velocity in a pipeline can be expressed as:

$$v = \sqrt{2gH}$$

This can be expressed in terms of H, and with a suitable coefficient this can then be used to find the head loss through pipe fittings, in the following form:

$$H = \frac{Cv^2}{2g}$$ **Formula 15.7**

where

H is the head loss through the fitting in metres
v is the velocity in m/s
g is the acceleration due to gravity (9·806 m/s^2)
C is a coefficient (see Table 15.3)

If the maximum velocity in a rising main is 3·0 m/s, the velocity head in the pipeline is then:

$$\frac{v^2}{2g} = \frac{3^2}{2 \times 9{\cdot}806} = 0{\cdot}46\ \text{m}$$

and the combined losses due to fittings and valves (see Table 15.3) may only be about 1·0 m. These head losses are therefore normally ignored unless the total manometric head is very low. The total head can be taken as the static head, plus friction losses as calculated earlier, plus something between 1·0 and 1·5 m to account for 'station losses'.

TABLE 15.3
VELOCITY HEAD IN FITTINGS AND VALVES

Fitting	*Coefficient (C)*
Bellmouth	0·10
Bend	0·25 to 0·50
Tee-junction	1·00
Sluice valve	0·15
Reflux valve (single flap)	0·70 to 2·00[a]

[a] The value of *C* for reflux valves increases as the velocity decreases.

THRUST BLOCKS AND ANCHORAGE

All pipelines which operate under pressure must be adequately anchored at bends, valves, tees, etc. Thrust blocks are usually constructed of *in situ* concrete, and their size will depend on the bearing capacity of the soil at the sides of the trench. Manufacturers of PVC pipes recommend the use of a thin membrane of bituminized paper, roofing felt or polythene film between the concrete and the pipe. This membrane need only be about 2 or 3 mm thick. Ductile iron pipelines incorporating anchor-type joints will not normally need concrete anchor blocks.

Where there is a change in direction in a pressure pipeline, there will be both dynamic and static thrust. In sewage rising mains it is preferable to make allowance for both types of thrust, although generally the static thrust will be the more important in view of the comparatively low velocities involved. The formula normally used is:

$$R = 0{\cdot}002\,(H + 0{\cdot}102\,v^2)\,A \sin \tfrac{1}{2}\phi \qquad \textbf{Formula 15.8}$$

where

R is the total thrust in kgf
H is the total hydrostatic head in metres
v is the velocity in m/s
A is the cross-sectional area of the pipeline in mm^2
ϕ is the angle of deviation in degrees

Table 7 of CP 310, 'Water Supply', gives imperial values of the end thrust and radial thrust at bends for pipes up to 300-mm diameter, working at a head of 70 m of water (100 lb per square inch). The values in the table can be multiplied by 1000 to give approximate values in kgf and, for heads of other than 70 m, values can be calculated *pro rata*.

TESTING

Rising mains should be tested with water or air on the same basis as a water supply main. For a water test, the test pressure may be from $1\frac{1}{2}$ times to twice the working pressure. A common allowance for leakage is of the order of 0·4 litres of water per millimetre of diameter, per kilometre length, for each 24 h, and for every 100 m head used in the test.

For an air test, the general *maximum* air test pressure is around 40 000 kg/m^2 and the main is considered to be in order if the pressure loss is negligible over a period of a few hours. Air testing can be dangerous, in view of the considerable kinetic energy which is stored in the pipeline during the test, and an air test should only be carried out by an experienced operative.

ECONOMICS

It is generally accepted that, taking into account the capital cost of the pumping equipment and the pipeline and the annual costs of pumping, the most economical diameter of rising main is when the velocity of flow at normal pumping rates is between 0·75 and 1·20 m/s and when the velocity at maximum rates of pumping is not more than about 1·8 m/s. The minimum internal diameter suitable for use with crude sewage is 100 mm.

The *approximate* velocity of flow in a rising main is:

$$v = \frac{353{\cdot}4\,Q}{d^2}$$ **Formula 15.9**

where
v is the velocity in m/s
Q is the pump discharge in m^3/h
d is the internal diameter of the pipeline in millimetres

The capital cost of a rising main naturally increases with the increase in diameter, but with that increase the velocity (and therefore the friction head) is reduced. Having obtained the friction head for normal rates of pumping through various alternative diameters, the power requirements can be calculated and the annual cost of pumping against each friction head can be estimated.

To obtain the most economic size of rising main, alternative calculations must be made and the capital and annual costs of each compared. To obtain the required range of velocities on a large scheme with pumps of varying outputs, it may be more economical to install duplicate rising mains. Where two sizes of pump are installed, it may then be more satisfactory to arrange for one main to take the flows from the smaller pumps, and for another to take the flows from the larger pumps. An emergency by-pass should then be included so that either main can act as a duty main if necessary.

References and Bibliography

Building Research Establishment

1. National Building Studies Special Report No. 32, *Simplified Tables of External Loads on Buried Pipelines*, 1962/63.
2. N.B.S Special Report No. 35, *Pipe Laying Principles*, 1964.
3. N.B.S. Special Report No. 37, *Loading Charts for the Design of Buried Rigid Pipes*, 1966.
4. N.B.S. Special Report No. 38, *High-strength Beddings for Unreinforced Concrete and Clayware Pipes*, 1967.
5. Digest No. 107, *Roof Drainage*, 1969.
6. Digest Nos. 124 and 125, *Small Underground Drains and Sewers I and II*, 1959.
7. Digest No. 6 (Second Series), *Drainage for Housing*, 1966.
8. Digest No. 80 (Second Series), *Soil and Waste Pipe Systems for Housing*, 1967.
9. Current Paper 58/68, *Specifying Concrete*, 1968.
10. Current Paper, Design Series 60, *Sanitary Services for Modern Housing*, 1966.
11. Current Paper, Engineering Series 23, *Pipeline Design—the Relationship between Structural Theory and Laying Practice*, 1964.
12. *Simplified Tables of External Loads on Buried Pipelines* (*Revised*), 1970.

Transport and Road Research Laboratory

13. Road Note No. 19, *The Design Thickness of Concrete Roads*, 1955.
14. Road Note No. 29, *A Guide to the Structural Design of Flexible and Rigid Pavements for New Roads*, 1970.
15. Road Note No. 35, *A Guide for Engineers to the Design of Storm Sewer Systems*, 1976.
16. Research Technical Paper No. 55, *The Design of Urban Sewer Systems—Research into the Relation Between the Rate of Rainfall and the Rate of Flow in Sewers*, 1962.
17. Report LR 110, *Subsoil Drainage and the Structural Design of Roads*, 1967.
18. Report LR 236, *The Depth of Rain Water on Road Surfaces*, 1968.
19. Report No. 35, *Impact Tests on Pipes Buried under Roads*, 1966.

Water Pollution Research Laboratory

20. Notes on Water Pollution No. 6, *Discharge of Sulphates into Concrete Sewers*, 1959.
21. Notes No. 8, *Sampling Natural Waters and Polluting Liquids*, 1960.
22. Notes No. 17, *Waste Waters from Farms*, 1962.
23. Notes No. 24, *Some Further Observations on Waste Waters from Farms*, 1964.
24. Notes No. 30, *Storm Sewage Investigations*, 1965.
25. Notes No. 32, *Formation of Sulphides in Sewers*, 1966.
26. Notes No. 39, *Pollution of Inland Waters by Oil*, 1967.
27. Reprint No. 509, 'Field Studies on the Flow and Composition of Storm Sewage', Davidson and Gameson.
28. Reprint No. 510, 'Storm Overflow Performance Studies using Crude Sewage', Ackers and Brewer.

Hydraulics Research Station

29. Research Paper No. 2, *Charts for the Hydraulic Design of Channels and Pipes*, 1969.
30. Research Paper No. 4, *Tables for the Hydraulic Design of Storm-drains, Sewers and Pipelines*, 1969.
31. Report Int. 158, *Oil Interceptors for Surface Drains—A Literature Survey*, I. M. Gardiner, May 1976.

Institution of Civil Engineers

32. Report No. VI, *The Organization of Civil Engineering Work*, 1946.
33. Report No. VII, *The Contract System in Civil Engineering*, 1946.
34. Report of the Joint Committee on Location of Underground Services, 1946/1963.
35. *An Introduction to Engineering Economics*, 1969.
36. *Safety in Sewers and at Sewage Works*, 1969.
37. *The Vibration of Concrete*, 1956.
38. *A Guide to Specifying Concrete*, 1967.
39. Standard Method of Measurement of Civil Engineering Quantities.

Technical Papers

40. Abernethy, L. L., 'Effect of Trench Conditions and Arch Encasement on Load-bearing Capacity of Vitrified Clay Pipes', Ohio State University, *Engineering Experimental Station Bulletin* No. 158, 1955.
41. Ackers, P. *et al.* 'Effects of Use on the Hydraulic Resistance of Drainage Conduits', *Proc. I.C.E.*, 1964.
42. Bartlett, R. E., 'The Structural Design of g.v.c. Pipelines', *J. Inst. P.H. Eng.*, London, 1967.
43. Best, R., 'An Aspect of Sewerage and Sewage Disposal', *Assn. of Rural District Council Surveyors*, 1968.
44. Braine, C. D. C., 'The Effect of Storage on Sewerage Design', *Proc. I.C.E.*, 1955.
45. Clarke, N. W. B., 'The Causes and Prevention of Failures in Salt-glazed Ware or Other Ceramic Pipelines', *Pub. Works and Mun. Services Congress*, London, 1958.
46. Clarke, N. W. B. and Young, O. C., 'Some Structural Aspects of the Design of Concrete Pipelines', *Proc. I.C.E.*, 1959.
47. Clarke, N. W. B. and Young, O. C., 'Loads on Underground Pipes caused by Vehicle Wheels', *Proc. I.C.E.*, 1962.
48. Clarke, N. W. B., 'The Wide Trench Condition and its Effect on the Loads Imposed on Rigid Underground Conduits', *Proc. I.C.E.*, 1963.
49. Copas, B. A., 'Storm Water Sewer Calculations', *J. Inst. P.H. Eng.*, 1957.
50. Evans, R. H., 'Applications of Prestressed Concrete to Water Supply and Drainage', *Proc. I.C.E.*, 1955.
51. Greer, D. M. and Moorhouse, D. C., 'Engineering-Geological Studies for Sewer Projects', *J. San. Eng. Div. A.S.C.E.*, 1968.
52. Hainsworth, I. H. *et al.*, 'Pilot Experiments to Determine the Loads Causing Failure of Sewer Pipes Under Roads', *Proc. I.C.E.*, 1967.
53. Hartley, C. J., 'Calculation of Foul Sewage Flows in Small Sewerage Schemes and Building Drainage Schemes', *J. Inst. P.H. Eng.*, 1968.
54. Hughes, B. P., 'Rational Concrete Mix Design', *Proc. I.C.E.*, 1960.
55. Marston, A. and Anderson, A. O., 'The Theory of Loads on Pipes in Ditches and Tests of Cement and Clay Drain Tile and Sewer Pipe', *Iowa Engineering Exp. Station Bulletin*, 31, 1913.
56. Mollinson, A. R., 'Road Surface Water Drainage', *J. Inst. Highway Eng.*, Vol. V, iv.
57. Ramseier, R. E. and Riek, G. C., 'Low Pressure Air Test for Sanitary Sewers'. *J. San. Eng. Div. A.S.C.E.*, 1964.
58. Rowe, K., 'Concrete Towards 2007', *J. Inst. Water Eng.*, 1967.
59. Sarginson, E. J. and Bourne, D. E., 'The Analysis of Urban Rainfall Run-off and Discharge', *J. Inst. Mun. E.*, March 1969.
60. Shaw, V. A., 'An Air-testing Procedure for the Determination of the Relative Tightness of Underground Sewers', Pretoria.
61. Storey, J. B., 'Some Factors Affecting the Construction of Extra Strength Pipe Sewers', *J. Inst. Mun. E.*, January 1966.
62. Walton, J. H., 'The Strength of Rigid Pipelines Buried in Embankments with Large Positive Projection', *J. Inst. P.H. Eng.*, 1966.

Textbooks, etc.

63. Barlow, T., *Hydraulics—Gauging of Sewage Flows, etc.*, Crosby Lockwood, 1926.
64. Bartlett, R. E., *Pumping Stations for Water and Sewage*, Applied Science, 1974.
65. Bartlett, R. E., *Surface Water Sewerage*, Applied Science, 1976.
66. Bartlett, R. E., *Wastewater Treatment*, Applied Science, 1971.
67. Clarke, N. W. B., *Buried Pipelines—A Manual of Structural Design and Installation*, Applied Science, 1968.
68. Crimp, S. and Bruges, W. E., *Tables and Diagrams for Use in Designing Sewers and Water Mains*, Mun. Publications Co. Ltd., 1969.
69. Escritt, L. B., *Sewerage and Sewage Disposal, Calculations and Design*, C. R. Books, 1967.
70. Escritt, L. B., *Sewer and Water-main Design Tables, British & Metric*, Applied Science, 1969.
71. Garner, J. F., *The Law of Sewers and Drains*, Shaw.
72. Haywood, L. M., *Survey Practice on Construction Sites*, Pitman, 1968.
73. Isaac, P. G. G., *Public Health Engineering*, Spon, 1953.
74. Lockyer, K. G., *An Introduction to Critical Path Analysis*, Pitman, 1964.
75. Spangler, M. G., *Soil Engineering*, International Textbook Co., 1960.
76. Twort, A. C., *The Supervision of Civil Engineering Construction*, Arnold, 1972.
77. Wise, A. F. E., *Drainage Pipework in Dwellings—Hydraulic Design and Performance*, H.M.S.O., 1967.
78. Woolley, E. L., *Drainage Details*, Northwood Industrial Pub., 1971.

Miscellaneous

79. *Australian Rainfall and Run-off*, Stormwater Standards Committee, Institution of Engineers, Australia.
80. *Clay Pipe Engineering Manual*, National Clay Pipe Institute, Crystal Lake, Illinois.
81. *Cast Iron Pipelines—Their Manufacture and Installation*, Stanton & Staveley, Nottingham.
82. *Concrete Pipelines*, American Concrete Pipe Association, Chicago.
83. *Data Sheets—Sewerage*, Municipal Engineering, London.
84. *Design and Construction of Sanitary and Storm Sewers*, Water Pollution Control Federation Manual of Practice No. 9; Amer. Society of Civil Engineers Manual of Engng. Practice No. 37.
85. *Design Charts for Vitrified Clay Pipelines*, Ellistown Pipes Ltd., 1968.
86. *Design Tables for Determining the Bedding Construction of Vitrified Clay Pipelines*, Clay Pipe Development Assn., London.
87. *Glazing Manual*. Glass and Glazing Federation, London.
88. *Granolithic Concrete*, British Granite and Whinstone Fed., London.
89. *Handbook of Drainage and Construction Products*, Armco International Corporation, Ohio.
90. *Jacking Concrete Pipes*, Technical Bulletin No. 5, Concrete Pipe Association, 1975.
91. *Modern Methods of Laying Concrete Pipes*, Concrete Pipe Association, London.
92. *Pipe Loading Calculation Charts*, South Wales Concrete Pipe Co. Ltd.
93. *Pitch-fibre Systems*, Pitch-fibre Pipe Association, 1974.
94. *PVC Pipelaying Manual*. The British Plastics Federation, London, 1967.
95. *RPM Reinforced Plastic Matrix Pipe*, Stanton and Staveley, 1977.
96. *Safe Load Tables for Concrete Blockwork*, Cement and Concrete Association, 1971.
97. *Sewage Pump Technical Manual*, Sigmund Pulsometer Pumps Ltd., Reading.
98. *Handbook on Structural Steel*, Properties and Safe Loads, British Constructional Steelwork Association, 1971.
99. *Technical Paper No.* 40, United States Weather Bureau.

A Some Relevant Government Publications

Local Government Act, 1972.
Public Health Acts, 1936 and 1961.
Public Health (Drainage of Trade Premises) Act, 1937.
Public Utilities Street Works Act, 1950.
Rivers (Prevention of Pollution) Acts, 1951 and 1961.
Rivers (Prevention of Pollution) (Scotland) Act, 1951.
Sewerage (Scotland) Act, 1968.
The Building Regulations, 1976.
The Building Regulations (Northern Ireland).
The Building Standards (Scotland) Regulations.
The Traffic Signs, Regulations and General Directions, 1964.
Town and Country Planning Act, 1971.
Water Act, 1973.
Department of Transport, Model Contract Document for Topographical Survey Contracts, 1978.
Ministry of Housing and Local Government, Working Party on the Design and Construction of Underground Pipe Sewers, Third Report, 1971.
Ibid, Working Party on Sewers and Water Mains, First Report, 1975.
Ibid, Technical Committee on Storm Overflows and the Disposal of Storm Sewage, Final Report, 1970.
Ministry of Works, Post-war Building Studies No. 26, Domestic Drainage, 1947.
Ministry of Public Building and Works, Advisory Leaflet No. 7, 'Concreting in Cold Weather'.
Ibid, Advisory Leaflet No. 66, 'Laying Flexible Drain and Sewer Pipes—such as Pitch-fibre and PVC', 1967.
Ibid, Advisory Leaflet No. 67, 'Building without Accidents', 1967.
Ibid, Statements DC4 and DC5, 'Dimensional Co-ordination for Building', 1968.
Ministry of Health, 'Accidents in Sewers—Report on the Precautions necessary for the Safety of Persons entering Sewers and Sewage Tanks', 1934.
Ministry of Agriculture, Fisheries and Food, 'Growmore' Leaflet No. 44, 'Mole Drainage for Heavy Land', 1967.
National Economic Development Office, 'Contracting in Civil Engineering since Banwell', 1968.

B Relevant British Standards Institution Publications

BSCP: 110, The Structural Use of Concrete.
Ibid., 112, Part 2, The Structural Use of Timber.
Ibid., 114, The Structural Use of Reinforced Concrete in Buildings, Part 2.
Ibid., 301, Building Drainage.
Ibid., 304, Sanitary Pipework above Ground.
Ibid., 2001, Site Investigations.
Ibid., 2003, Earthworks.
Ibid., 2004, Foundations.
Ibid., 2005, Sewerage.
Ibid., 2010, Pipelines—Installation of Pipelines in Land.
BSS: 4, Structural Steel Sections—Hot Rolled Sections, Part 1.
Ibid., 12, Portland Cement (Ordinary and Rapid Hardening).
Ibid., 65 and 540, Clay Drain and Sewer Pipes, including Surface Water Pipes and Fittings.
Ibid., 78, Cast Iron Spigot and Socket Pipes for Water, Sewage, etc.: Part 2, *Fittings*.
Ibid., 308, Engineering Drawing Office Practice.
Ibid., 350, Conversion Factors and Tables.
Ibid., 437, Cast Iron Spigot and Socket Drain Pipes.
Ibid., 449, Part 2, The Use of Structural Steel in Building.
Ibid., 486, Asbestos-Cement Pressure Pipes.
Ibid., 497, Cast Manhole Covers and Road Gully Gratings and Frames for Drainage Purposes.
Ibid., 534, Steel Spigot and Socket Pipes and Specials for Water, Gas and Sewage.
Ibid., 539, Dimensions of Fittings for use with Clay Drain and Sewer Pipes.
Ibid., 556, Concrete Cylindrical Pipes and Fittings.
Ibid., 587, Motor Starters and Controllers.
Ibid., 599, Methods of Testing Pumps.
Ibid., 882 and 1201, Aggregates from Natural Sources for Concrete (including Granolithic).
Ibid., 1143, Salt-glazed Pipes with Chemically Resistant Properties.
Ibid., 1192, Building Drawing Practice.
Ibid., 1194, Concrete Porous Pipes for Under-drainage.
Ibid., 1196, Clayware Field Drain Pipes.
Ibid., 1211, Spun Iron Pipes for Water, Gas and Sewage.
Ibid., 1247, Manhole Stepirons.
Ibid., 1347, Architects Engineers and Surveyors Scales—Part 3—Metric Scales.
Ibid., 1377, Testing Soils for Engineering Purposes.
Ibid., 1957, The Presentation of Numerical Values.
Ibid., 2028, and 1364, Precast Concrete Blocks.
Ibid., 2035, Cast Iron Flanged Pipes and Flanged Fittings.
Ibid., 2494, Rubber Joint Rings.
Ibid., 2760, Pitch-impregnated Fibre Drain and Sewer Pipes.
Ibid., 3429, Sizes of Drawing Sheets.
Ibid., 3464, Cast Iron Wedge and Double Disk Gate Valves for General Purposes.
Ibid., 3505, uPVC Pipe for Cold Water Services.
Ibid., 3506, uPVC Pipe for Industrial Uses.
Ibid., 3656, Asbestos-cement Pipes and Fittings for Sewerage and Drainage.
Ibid., 3763, International System (SI) Units.
Ibid., 3921, Part 2, Bricks and Blocks of Fired Brickearth, Clay or Shale.

Ibid., 4000, Specification for Sizes of Paper and Boards.
Ibid., 4011, Co-ordination of Dimensions in Building—Basic Sizes for Building Components and Assemblies.
Ibid., 4027, Sulphate-resisting Portland Cement.
Ibid., 4101, Concrete Unreinforced Tubes and Fittings with Ogee Joints for Surface Water Drainage.
Ibid., 4318, Preferred Metric Basic Sizes for Engineering.
Ibid., 4330, Recommendations for Controlling Dimensions (Metric Units).
Ibid., 4449, Hot Rolled Steel Bars for the Reinforcement of Concrete.
Ibid., 4461, Cold Worked Steel Bars for the Reinforcement of Concrete.
Ibid., 4466, Building Dimensions and Scheduling of Bars for the Reinforcement of Concrete.
Ibid., 4471, Dimensions of Softwood.
Ibid., 4482, Hard Drawn Mild Steel Wire for the Reinforcement of Concrete.
Ibid., 4483, Steel Fabric for the Reinforcement of Concrete.
Ibid., 4622, Grey Iron Pipes and Fittings.
Ibid., 4625, Prestressed Concrete Pipes (Including Fittings).
Ibid., 4660, Unplasticized PVC Underground Drain Pipe and Fittings.
Ibid., 4772, Ductile Iron Pipes and Fittings.
Ibid., 5337, The Structural Use of Concrete for Retaining Aqueous Liquids (Formerly CP 2007).
Ibid., 5450, Sizes of Hardwoods and Methods of Measurement.

C Definitions and Abbreviations

DEFINITIONS (see also CP 2005)

Back-drop manhole: A manhole built at a junction of two sewers, where one sewer joins the other at a higher level and the sewage passes through a vertical or inclined shaft to the lower level.

Benching: A surface at the base of a chamber with the dual purpose of confining the flow of sewage to avoid the accumulation of deposits and of providing a safe working surface.

Catchment area: The area of a watershed discharging to a sewer, river or lake.

Combined sewer: A sewer designed to carry both foul sewage and surface water.

Crude sewage: Sewage which has received no treatment.

Datum: A plane of reference for a system of levels.

Dilution: The ratio of the volume of a stream to the volume of sewage or effluent discharging to it.

Dry weather flow: The rate of flow of sewage, together with infiltration, if any, in a sewer in dry weather—measured after a period of seven consecutive days of dry weather during which the rainfall has not exceeded 0·25 mm.

Ejector: A means of raising sewage by admitting it through a valve to a closed vessel, and then ejecting it through another valve by admitting compressed air into the vessel.

Flushing manhole: A manhole provided with a penstock so that water or sewage can be accumulated and then discharged to flush the sewer downstream.

Foul sewage: Any water contaminated by domestic waste or trade effluents.

Gradient: The inclination of the invert of a pipeline expressed as a fall in a given length.

Gravity sewer: A sewer in which the sewage runs from one end to the other on a descending gradient, and where pumping is not required.

Hydraulic gradient: The surface slope of a liquid in a pipeline. This is generally taken as parallel to the invert in a gravity sewer.

Infiltration: The entry of groundwater into a sewer from the surrounding soil.

Invert: The lowest point of the internal cross section of a sewer or channel.

Inverted siphon: A portion of a pipe or conduit in which sewage flows under pressure, due to the sewer dropping below the hydraulic gradient and then rising again.

Kilogramme: See BS 3763.

Lamp-hole: A small shaft constructed of pipes for the purpose of lowering a lamp into the sewer to facilitate inspection. Lamp-holes are now rarely constructed on new sewers.

Manhole: A chamber constructed on a sewer so as to provide access thereto for inspection, testing or the clearance of obstruction.

Metre: See BS 3763.

Outfall sewer: The final sewer of a system which carries the sewage to a treatment works or to a point of discharge.

Pumping station: An installation of pumps to lift sewage from a low level to a higher level.

Rainfall: Precipitation in any form, such as rain, snow, hail, etc. The rate of rainfall is measured in millimetres per hour.

Run-off: That part of rainfall which flows off the surface to reach a sewer or river.

Rising main: A pipeline carrying the discharge from a pump (or pumps) which is running full and at a pressure greater than atmospheric.

Screen: A device with openings designed to retain coarse solids from sewage. The openings are usually rectangular slots between evenly spaced bars.

Screenings: Material removed from sewage at a screen.

Separate sewer: A sewer designed to carry foul sewage only.

Sewage: Water-borne human, domestic and farm waste. It may include trade effluent, subsoil or surface water.

Sewage treatment: An artificial process to which sewage is subjected to remove or alter its constituents to render it less offensive or dangerous.

Sewer: A pipeline to carry sewage or other wastes, and not normally flowing full.

Sewerage: A system of sewers and ancillary works to convey sewage from its point of origin to a treatment works or other place of disposal.

Sludge: Accumulated suspended solids from sewage deposited in pipes or tanks, mixed with water to form a semi-liquid substance.
Soffit: The highest point of the internal cross-section of a sewer.
Storm sewage: Foul sewage diluted with surface water.
Storm sewage overflow: A weir, siphon or other device on a combined or partially separate sewerage system, introduced for the purpose of relieving the system of flows in excess of a selected rate, so that the size of the sewers downstream of the overflow can be kept within economical limits, the excess flow being discharged to a convenient watercourse.
Surface water: Natural water from the ground surface, paved areas and roofs.
Trade effluent: The fluid discharge, with or without matters in suspension, resulting wholly or in part from any manufacturing process, and including farm and research institution effluents.
Trunk sewer: A main sewer which takes the flow from a number of branch sewers, and serves as the main carrier of sewage for a large area.

ABBREVIATIONS

Ampere(s)	A	Kilovolt-ampere	kVA
Cubic metre(s)	m^3	Kilowatt(s)	kW
Cubic metres per second	m^3/s or cumec	Litre(s)	l
Day(s)	d	Metre(s)	m
Dry weather flow	d.w.f.	Metres per second	m/s
Gravitational acceleration ($9{\cdot}806\ m/s^2$)	g	Millimetre(s)	mm
Hectare ($10^4\ m^2$)	ha	Minute(s)	min
Hour(s)	h	Newton(s)	N
Hertz (frequency)	Hz	Revolutions per minute	rev/min
Kilogramme(s)	kg	Second(s)	s
Kilogramme-force	kgf	Square metre(s)	m^2
Kilometre(s)	km	Year (annum)	a

D Weights of Materials

CUBIC MEASURE

(*See also Chapter* 4 *and BS* 648)

	Weight in kg/m^3
Ashes and coke (loose)	640
Broken stone (10 mm to sand)	1 250
Ballast (all-in)	1 800
Brickwork, commons	2 000
Brickwork, engineering	2 400
Clay, puddled	2 160
Cement, Portland	1 440
Concrete, unreinforced	2 300
Concrete, reinforced (2 % steel)	2 420
Concrete, foamed slag	1 280
Clinker	800
Cast iron	7 200
Coal (loose)	880
Gravel (10 mm to sand)	1 400
Hardcore (average)	1 920
Lead	11 400
Limestone	2 640
Macadam	2 400
Oil, fuel	900
Petrol	670
Pitch and tar	1 280
Sand, average loose weight	1 600
Stone, York	2 400
Stone, Bath	2 240
Stone, rubble (packed)	2 240
Sandstone	2 240
Snow, newly fallen	100
Snow, compacted (up to)	800
Timber, ash (approx.)	800
Timber, pine (approx.)	575
Timber, oak (approx.)	850

SQUARE MEASURE

	Weight in kg/m^2
Asphalt, rock or mastic, per 25-mm thickness	53
Breeze blockwork (75 mm thick)	100
Brickwork (225 mm thick)	440
Glass, per 5-mm thickness	12·5
Granolithic finish, per 25-mm thickness	60
Sand/cement screed, per 25-mm thickness	58

E Miscellaneous Formulae and Tables

USEFUL FORMULAE

1. *Triangles:* (i) $a^2 = b^2 + c^2 - 2bc \cos A$

(ii) $\dfrac{a}{\sin A} = \dfrac{b}{\sin B} = \dfrac{c}{\sin C}$

(iii) Area $= \sqrt{s(s-a)(s-b)(s-c)}$

where $s = \frac{1}{2}(a + b + c)$

(iv) Area $= \dfrac{ab}{2}\sin C = \dfrac{ac}{2}\sin B = \dfrac{bc}{2}\sin A$

2. *Trapezium:* Area $= \dfrac{b + b'}{2}h$

3. *Sector of circle:* Area $= \dfrac{\pi R^2}{360}\alpha$

where α is the included angle

4. *Segment of circle:* Area $= \dfrac{\pi R^2}{360}\alpha - \dfrac{C}{2}(R - S)$

where: R is radius
C is chord
S is depth of segment

5. *Pyramid:* Volume $= 1/3\,Bh$

6. *Circular cone:* Volume $= \pi r^2 h/3$.

RELATIVE DISCHARGING POWER OF SMALL DIAMETER PIPES

Nominal internal diameter, mm					
150	*100*	*75*	*50*	*25*	*12·5*
1	2	5	15	88	498
—	1	2	5	32	181
—	—	1	2	15	88
—	—	—	1	5	32
—	—	—	—	1	5

TABLE OF ARCS, CHORDS AND AREAS OF SEGMENT

Where radius (R) is 1·0

Angle in degrees	*Arc*	*Chord, C*	*Area of segment*
1	0·0175	0·0175	0·00000
2	0·0349	0·0349	0·00000
3	0·0524	0·0524	0·00001
4	0·0698	0·0698	0·00003
5	0·0873	0·0873	0·00006
6	0·1047	0·1047	0·00010
7	0·1222	0·1221	0·00015
8	0·1396	0·1395	0·00023
9	0·1571	0·1569	0·00032
10	0·1745	0·1743	0·00044
11	0·1920	0·1917	0·00059
12	0·2094	0·2091	0·00076
13	0·2269	0·2264	0·00097
14	0·2443	0·2437	0·00121
15	0·2618	0·2611	0·00149
16	0·2793	0·2783	0·00181
17	0·2967	0·2956	0·00217
18	0·3142	0·3129	0·00257
19	0·3316	0·3301	0·00302
20	0·3491	0·3473	0·00352
21	0·3665	0·3645	0·00408
22	0·3840	0·3816	0·00468
23	0·4014	0·3987	0·00535
24	0·4189	0·4158	0·00607
25	0·4363	0·4329	0·00686
26	0·4538	0·4499	0·00771
27	0·4712	0·4669	0·00862
28	0·4887	0·4838	0·00961
29	0·5061	0·5008	0·01067
30	0·5236	0·5176	0·01180
31	0·5411	0·5345	0·01301
32	0·5585	0·5512	0·01429
33	0·5760	0·5680	0·01566
34	0·5934	0·5847	0·01711
35	0·6109	0·6014	0·01864
36	0·6283	0·6180	0·02027
37	0·6458	0·6346	0·02198
38	0·6632	0·6511	0·02378
39	0·6807	0·6676	0·02568
40	0·6981	0·6840	0·02767
41	0·7156	0·7004	0·02976
42	0·7330	0·7167	0·03195
43	0·7505	0·7330	0·03425
44	0·7679	0·7492	0·03614
45	0·7854	0·7654	0·03915
46	0·8029	0·7815	0·04176

TABLE OF ARCS, CHORDS AND AREAS OF SEGMENT *(contd.)*

Angle in degrees	*Arc*	*Chord, C*	*Area of segment*
47	0·820 3	0·797 5	0·044 48
48	0·837 8	0·813 5	0·047 31
49	0·855 2	0·829 4	0·050 25
50	0·872 7	0·845 2	0·053 31
51	0·890 1	0·861 0	0·056 49
52	0·907 6	0·876 7	0·059 78
53	0·925 0	0·892 4	0·063 19
54	0·942 5	0·908 0	0·066 73
55	0·959 9	0·923 5	0·070 39
56	0·977 4	0·938 9	0·074 17
57	0·994 8	0·954 3	0·078 08
58	1·012 3	0·969 6	0·082 12
59	1·029 7	0·984 8	0·086 29
60	1·047 2	1·000 0	0·090 59
61	1·064 7	1·015 1	0·095 02
62	1·082 1	1·030 1	0·099 58
63	1·099 6	1·045 0	0·104 28
64	1·117 0	1·059 8	0·109 11
65	1·134 5	1·074 6	0·114 08
66	1·151 9	1·089 3	0·119 19
67	1·169 4	1·103 9	0·124 43
68	1·186 8	1·118 4	0·129 82
69	1·204 3	1·132 8	0·135 35
70	1·221 7	1·147 2	0·141 02
71	1·239 2	1·161 4	0·146 83
72	1·256 6	1·175 6	0·152 79
73	1·274 1	1·189 6	0·158 89
74	1·291 5	1·203 6	0·165 14
75	1·309 0	1·217 5	0·171 54
76	1·326 5	1·231 3	0·178 08
77	1·343 9	1·245 0	0·184 77
78	1·361 4	1·258 6	0·191 60
79	1·378 8	1·272 2	0·198 59
80	1·396 3	1·285 6	0·205 73
81	1·413 7	1·298 9	0·213 01
82	1·431 2	1·312 1	0·220 45
83	1·448 6	1·325 2	0·228 04
84	1·466 1	1·338 3	0·235 78
85	1·483 5	1·351 2	0·243 67
86	1·501 0	1·364 0	0·251 71
87	1·518 4	1·376 7	0·259 90
88	1·535 9	1·389 3	0·268 25
89	1·553 3	1·401 8	0·276 75
90	1·570 8	1·414 2	0·285 40

TABLE OF CIRCUMFERENCES AND AREAS OF CIRCLES

Diameter, d	*Circumference, πd*	*Area, $\frac{\pi d^2}{4}$*
1	3·142	0·785
2	6·283	3·142
3	9·425	7·069
4	12·566	12·566
5	15·708	19·635
6	18·850	28·274
7	21·991	38·484
8	25·133	50·265
9	28·274	63·617
10	31·416	78·540
11	34·558	95·033
12	37·699	113·097
13	40·841	132·732
14	43·982	153·938
15	47·124	176·715
16	50·265	201·062
17	53·407	226·980
18	56·549	254·469
19	59·690	283·529
20	62·832	314·159
21	65·973	346·361
22	69·115	380·133
23	72·257	415·476
24	75·398	452·389
25	78·540	490·874
26	81·681	530·929
27	84·823	572·555
28	87·965	615·752
29	91·106	660·520
30	94·248	706·858
31	97·389	754·768
32	100·531	804·248
33	103·673	855·299
34	106·814	907·920
35	109·956	962·113
36	113·097	1 017·88
37	116·239	1 075·21
38	119·381	1 134·11
39	122·522	1 194·59
40	125·66	1 256·64
41	128·81	1 320·25
42	131·95	1 385·44
43	135·09	1 452·20
44	138·23	1 520·53

TABLE OF CIRCUMFERENCES AND AREAS OF CIRCLES *(contd.)*

Diameter, d	*Circumference, πd*	*Area, $\frac{\pi d^2}{4}$*
45	141·37	1 590·43
46	144·51	1 661·90
47	147·65	1 734·94
48	150·80	1 809·56
49	153·94	1 885·74
50	157·08	1 963·50
51	160·22	2 042·82
52	163·36	2 123·72
53	166·50	2 206·18
54	169·65	2 290·22
55	172·79	2 375·83
56	175·93	2 463·01
57	179·07	2 551·76
58	182·21	2 642·08
59	185·35	2 733·97
60	188·50	2 827·43
61	191·64	2 922·47
62	194·78	3 019·07
63	197·92	3 117·25
64	201·06	3 216·99
65	204·20	3 318·31
66	207·35	3 421·19
67	210·49	3 525·65
68	213·63	3 631·68
69	216·77	3 739·28
70	219·91	3 848·45
71	223·05	3 959·19
72	226·19	4 071·50
73	229·34	4 185·39
74	232·48	4 300·84
75	235·62	4 417·86
76	238·76	4 536·46
77	241·90	4 656·63
78	245·04	4 778·36
79	248·19	4 901·67
80	251·33	5·026·55
81	254·47	5 153·00
82	257·61	5 281·02
83	260·75	5 410·61
84	263·89	5 541·77
85	267·04	5 674·50
86	270·18	5 808·80
87	273·32	5 944·68
88	276·46	6 082·12
89	279·60	6 221·14
90	282·74	6 361·73

TABLE OF CIRCUMFERENCES AND AREAS OF CIRCLES *(contd.)*

Diameter, d	*Circumference, πd*	*Area, $\frac{\pi d^2}{4}$*
91	285·88	6 503·88
92	289·03	6 647·61
93	292·17	6 792·91
94	295·31	6 939·78
95	298·45	7 088·22
96	301·59	7 238·23
97	304·73	7 389·81
98	307·88	7 542·96
99	311·02	7 697·69
100	314·16	7 853·98
101	317·30	8 011·85
102	320·44	8 171·28
103	323·58	8 332·29
104	326·73	8 494·87
105	329·87	8 659·01
106	333·01	8 824·73
107	336·15	8 992·02
108	339·29	9 160·88
109	342·43	9 331·32
110	345·58	9 503·32
111	348·72	9 676·89
112	351·86	9 852·03
113	355·00	10 028·7
114	358·14	10 207·0
115	361·28	10 386·9
116	364·42	10 568·3
117	367·57	10 751·3
118	370·71	10 935·9
119	373·85	11 122·0
120	376·99	11 309·7

CONVERSION OF FLOW RATES

(N.B.—1 000 l/s = 1 cumec)

Litres per second, l/s	Cubic metres per hour, m^3/h	Cubic metres per hour, m^3/h	Litres per second, l/s
1	3·6	1	0·277
2	7·2	2	0·555
3	10·8	3	0·833
4	14·4	4	1·111
5	18·0	5	1·388
6	21·6	6	1·666
7	25·2	7	1·944
8	28·8	8	2·222
9	32·4	9	2·500
10	36·0	10	2·777
12	43·2	12	3·333
14	50·4	14	3·888
16	57·6	16	4·444
18	64·8	18	5·000
20	72·0	20	5·555
25	90·0	25	6·944
30	108·0	30	8·333
35	126·0	35	9·722
40	144·0	40	11·111
45	162·0	45	12·500
50	180·0	50	13·888
55	198·0	55	15·277
60	216·0	60	16·666
65	234·0	65	18·055
70	252·0	70	19·444
75	270·0	75	20·833
80	288·0	80	22·222
85	306·0	85	23·611
90	324·0	90	25·000
95	342·0	95	26·388
100	360·0	100	27·777
110	396·0	110	30·555
120	432·0	120	33·333
130	468·0	130	36·111
140	504·0	140	38·888
150	540·0	150	41·666
160	576·0	160	44·444
170	612·0	170	47·222
180	648·0	180	50·000
190	684·0	190	52·777
200	720·0	200	55·555
250	900·0	250	69·444
300	1 080·0	300	83·333
350	1 260·0	350	97·222
400	1 440·0	400	111·111
450	1 620·0	450	125·000
500	1 800·0	500	138·888
600	2 160·0	600	166·666
700	2 520·0	700	194·444
800	2 880·0	800	222·222
900	3 240·0	900	250·000
1 000	3 600·0	1 000	277·777

REPAYMENT OF LOANS (ANNUAL PAYMENTS OF CAPITAL AND INTEREST PER £100)

Interest %	*10 years*	*15 years*	*20 years*	*30 years*	*40 years*	*60 years*
6	13·59	10·30	8·72	7·26	6·65	6·19
7	14·24	10·98	9·44	8·06	7·50	7·12
8	14·90	11·68	10·18	8·88	8·39	8·08
9	15·58	12·41	10·96	9·73	9·30	9·05
10	16·27	13·15	11·75	10·61	10·23	10·03
11	16·98	13·91	12·56	11·50	11·17	11·02
12	17·70	14·68	13·39	12·42	12·13	12·01

F Conversion Factors

Metric unit	*Imperial equivalent*	*Imperial unit*	*Metric equivalent*
1 m	3·281 feet	1 ft	0·304 8 m
1 m	1·094 yards	1 yd	0·914 4 m
1 m	0·049 7 chains	1 chain	20·116 8 m
1 m	0·547 fathoms	1 fathom	1·829 m
1 mm	0·039 37 inches	1 inch	25·40 mm
1 km	0·621 4 miles	1 mile	1·609 km
1 m^2	10·76 sq ft	1 sq ft	0·092 90 m^2
1 m^2	1·196 sq yd	1 sq yd	0·836 1 m^2
1 mm^2	1·550 $\times 10^{-3}$ sq in	1 sq in	645·2 mm^2
1 km^2	0·386 1 sq miles	1 sq mile	2·590 km^2
1 ha	2·471 acres	1 acre	0·404 7 ha
1 m^3	35·31 cu ft	1 cu ft	0·028 32 m^3
1 m^3	1·308 cu yd	1 cu yd	0·764 6 m^3
1 litre	0·220 0 Imp gallons	1 Imp gall	4·546 litre
1 litre	0·264 2 US gallons	1 US gall	3·785 3 litre
1 litre	1·760 pints	1 pint	0·568 litre
1 l/s	13·20 gall per min	1 g.p.m.	0·075 77 l/s
1 m^3/h	3·675 Imp g.p.m.	1 g.p.m.	0·272 m^3/h
1 cumec	35·31 cusec	1 cusec	0·028 32 cumec
1 cumec	19·01 m.g.d.	1 m.g.d.	0·052 61 cumec
1 m^3/m^2	183·9 gall/sq yd	1 gall/sq yd	5·437 $\times 10^{-3}$ m^3/m^2
1 kg	2·205 pounds	1 pound	0·453 6 kg
1 kg	0·019 7 cwt	1 cwt	50·802 kg
1 kg/m^2	29·5 ounces per sq yd	1 oz per sq yd	33·90 $\times 10^{-3}$ kg/m^2
1 kg/m^3	0·062 43 pounds per cu ft	1 lb per cu ft	16·02 kg/m^3
1 tonne	0·984 2 ton	1 ton	1·016 tonne
1 N	0·224 8 lbf	1 lbf	4·448 N
1 N/m^2	0·145 $\times 10^{-3}$ p.s.i.	1 p.s.i.	6 894·76 N/m^2
1 N/m^2	0·004 01 inches w.g.	1 inch w.g.	249·089 N/m^2
1 kg/m^2	0·001 4 lb per sq in	1 p.s.i.	703 kg/m^2
1 kW	1·341 h.p.	1 h.p.	0·745 7 kW
1 deg C	1·8 deg F	1 deg F	0·555 deg C
1 m/s	3·281 feet per sec	1 ft per sec	0·304 8 m/s
1 km/h	0·621 4 m.p.h.	1 m.p.h.	1·609 km/h

100 N/m^2 = 1 mb
100 kN/m^2 = 1 bar
1 000 mb (1 bar) = 29·53 in mercury (at standard conditions)
°C = 5/9(°F − 32)

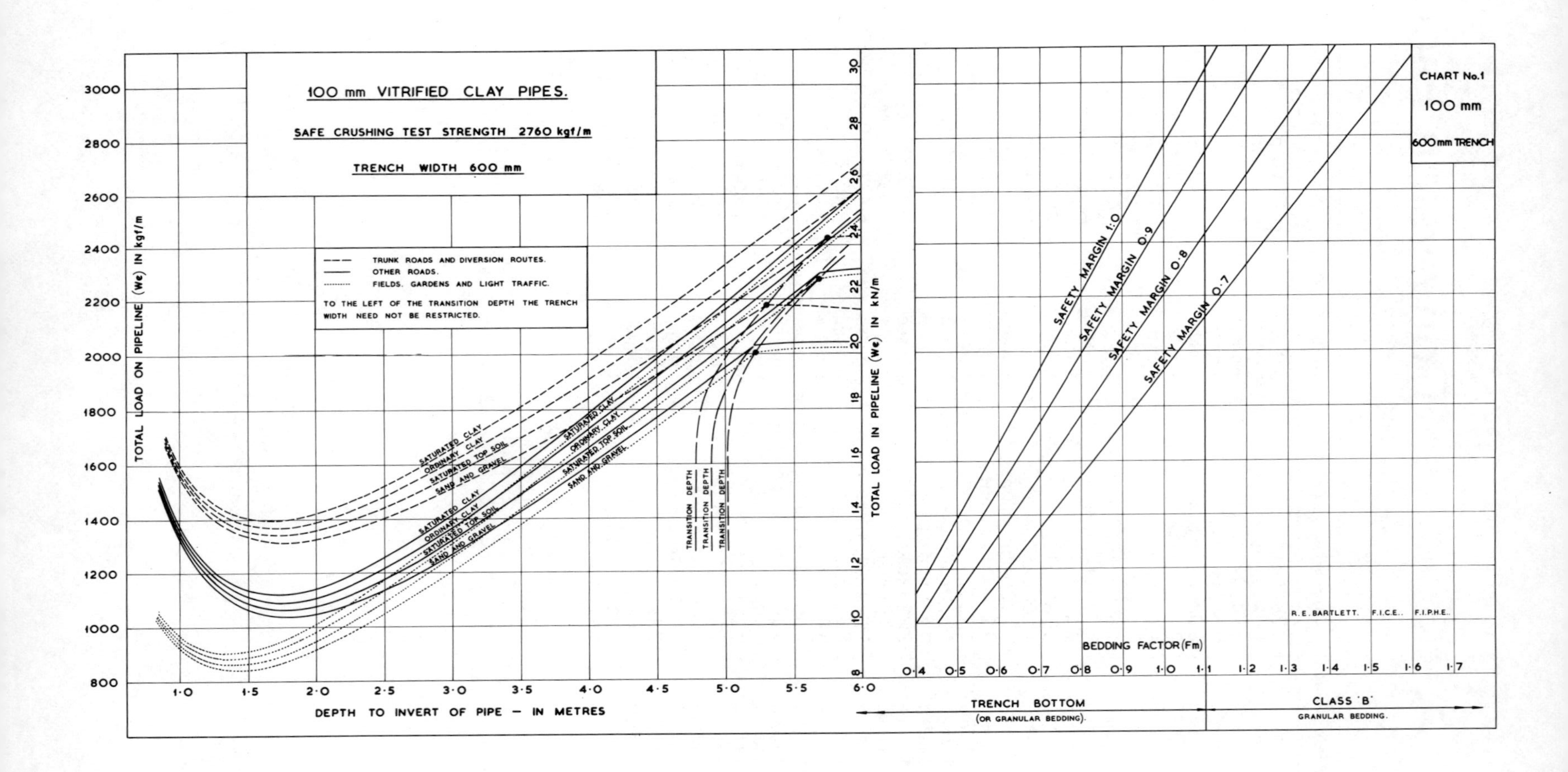

100 mm VITRIFIED CLAY PIPES.
SAFE CRUSHING TEST STRENGTH 2760 kgf/m
TRENCH WIDTH 600 mm
TRUNK ROADS AND DIVERSION ROUTES.
OTHER ROADS.
FIELDS. GARDENS AND LIGHT TRAFFIC.
TO THE LEFT OF THE TRANSITION DEPTH THE TRENCH WIDTH NEED NOT BE RESTRICTED.
SATURATED CLAY
ORDINARY CLAY
SATURATED TOP SOIL
SAND AND GRAVEL
TRANSITION DEPTH
TOTAL LOAD ON PIPELINE (We) IN kgf/m
3000
2800
2600
2400
2200
2000
1800
1600
1400
1200
1000
800
1·0
1·5
2·0
2·5
3·0
3·5
4·0
4·5
5·0
5·5
6·0
DEPTH TO INVERT OF PIPE — IN METRES
TOTAL LOAD IN PIPELINE (We) IN kN/m
8
10
12
14
16
18
20
22
24
26
28
30
SAFETY MARGIN 1·0
SAFETY MARGIN 0·9
SAFETY MARGIN 0·8
SAFETY MARGIN 0·7
CHART No.1
100 mm
600 mm TRENCH
R. E. BARTLETT. F.I.C.E.. F.I.P.H.E..
BEDDING FACTOR (Fm)
0·4
0·5
0·6
0·7
0·8
0·9
1·0
1·1
1·2
1·3
1·4
1·5
1·6
1·7
TRENCH BOTTOM
(OR GRANULAR BEDDING).
CLASS 'B'
GRANULAR BEDDING.

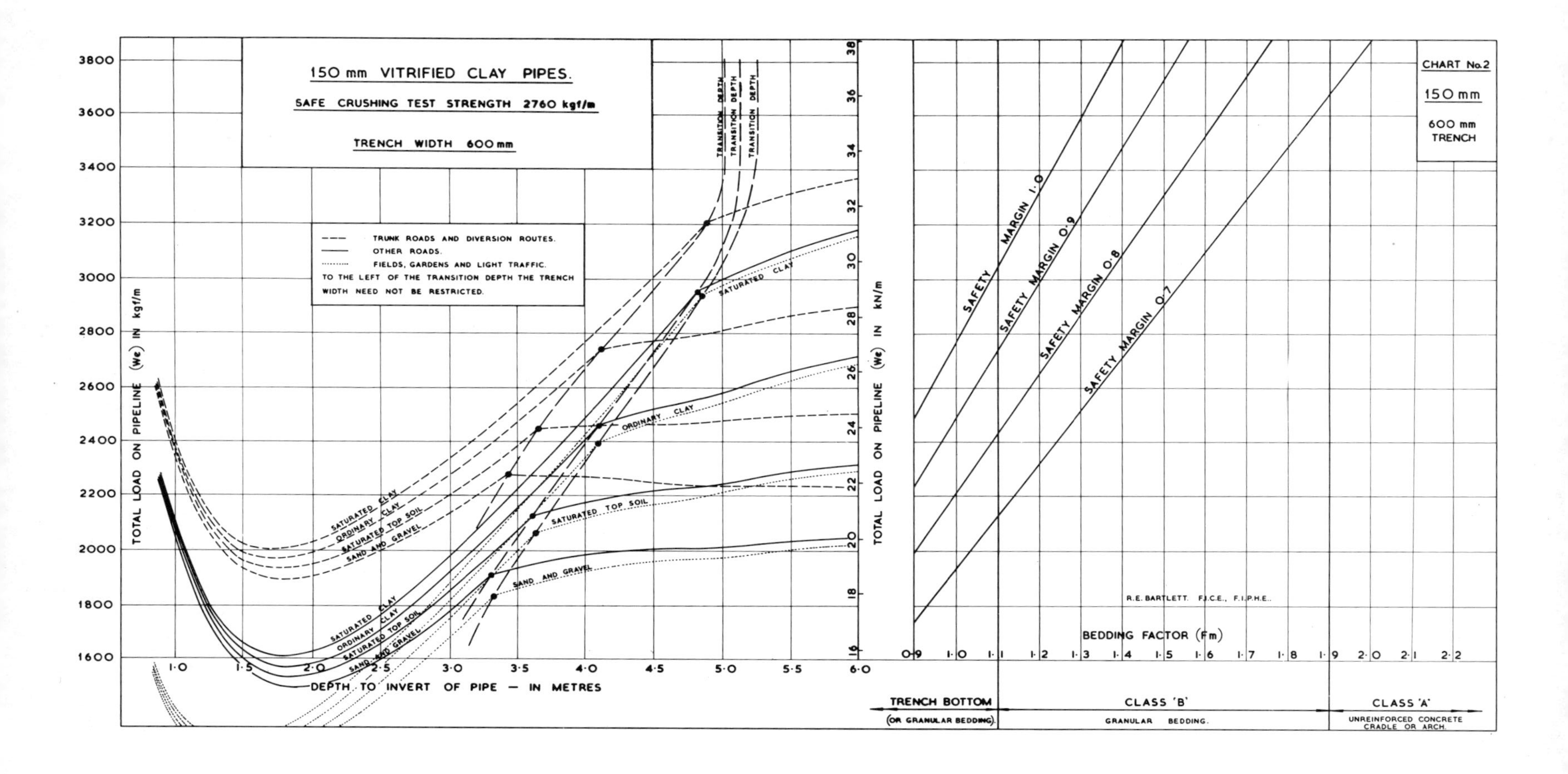
150 mm VITRIFIED CLAY PIPES.
SAFE CRUSHING TEST STRENGTH 2760 kgf/m
TRENCH WIDTH 600 mm
CHART No.2
150 mm
600 mm TRENCH
TRUNK ROADS AND DIVERSION ROUTES.
OTHER ROADS.
FIELDS, GARDENS AND LIGHT TRAFFIC.
TO THE LEFT OF THE TRANSITION DEPTH THE TRENCH WIDTH NEED NOT BE RESTRICTED.
TOTAL LOAD ON PIPELINE (We) IN kgf/m
TOTAL LOAD ON PIPELINE (We) IN kN/m
DEPTH TO INVERT OF PIPE — IN METRES
TRANSITION DEPTH
SATURATED CLAY
ORDINARY CLAY
SATURATED TOP SOIL
SAND AND GRAVEL
SAFETY MARGIN 1·0
SAFETY MARGIN 0·9
SAFETY MARGIN 0·8
SAFETY MARGIN 0·7
BEDDING FACTOR (Fm)
R.E. BARTLETT. F.I.C.E., F.I.P.H.E.
TRENCH BOTTOM
(OR GRANULAR BEDDING)
CLASS 'B'
GRANULAR BEDDING.
CLASS 'A'
UNREINFORCED CONCRETE CRADLE OR ARCH.

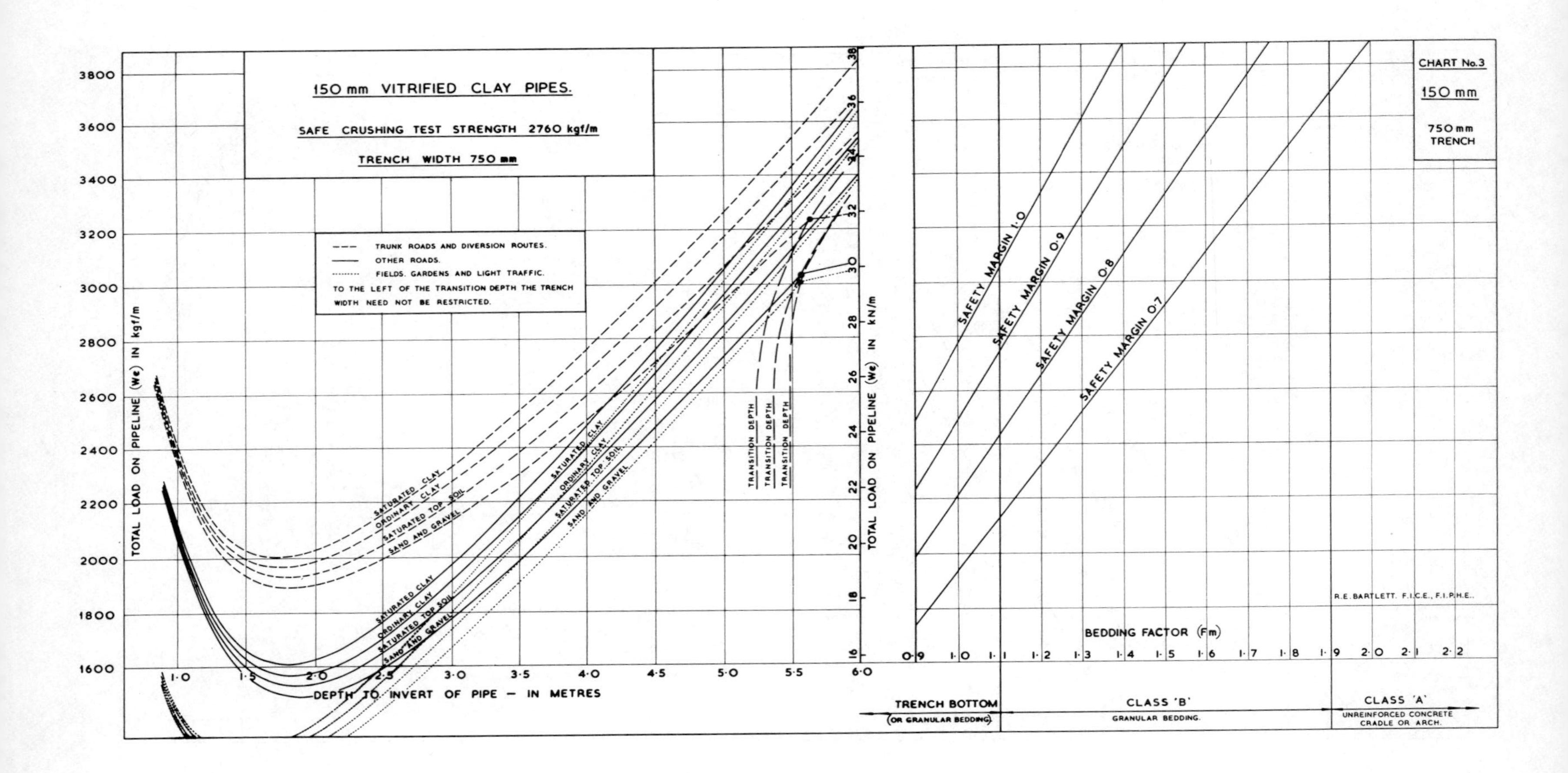

150 mm VITRIFIED CLAY PIPES.
SAFE CRUSHING TEST STRENGTH 2760 kgf/m
TRENCH WIDTH 750 mm
CHART No. 3
150 mm
750 mm TRENCH
TRUNK ROADS AND DIVERSION ROUTES.
OTHER ROADS.
FIELDS, GARDENS AND LIGHT TRAFFIC.
TO THE LEFT OF THE TRANSITION DEPTH THE TRENCH WIDTH NEED NOT BE RESTRICTED.
SATURATED CLAY
ORDINARY CLAY
SATURATED TOP SOIL
SAND AND GRAVEL
TRANSITION DEPTH
SAFETY MARGIN 1·0
SAFETY MARGIN 0·9
SAFETY MARGIN 0·8
SAFETY MARGIN 0·7
TOTAL LOAD ON PIPELINE (We) IN kgf/m
3800
3600
3400
3200
3000
2800
2600
2400
2200
2000
1800
1600
TOTAL LOAD ON PIPELINE (We) IN kN/m
38
36
34
32
30
28
26
24
22
20
18
16
1·0
1·5
2·0
2·5
3·0
3·5
4·0
4·5
5·0
5·5
6·0
DEPTH TO INVERT OF PIPE — IN METRES
BEDDING FACTOR (Fm)
0·9 1·0 1·1 1·2 1·3 1·4 1·5 1·6 1·7 1·8 1·9 2·0 2·1 2·2
R.E. BARTLETT. F.I.C.E., F.I.P.H.E.
TRENCH BOTTOM
(OR GRANULAR BEDDING)
CLASS 'B'
GRANULAR BEDDING.
CLASS 'A'
UNREINFORCED CONCRETE CRADLE OR ARCH.

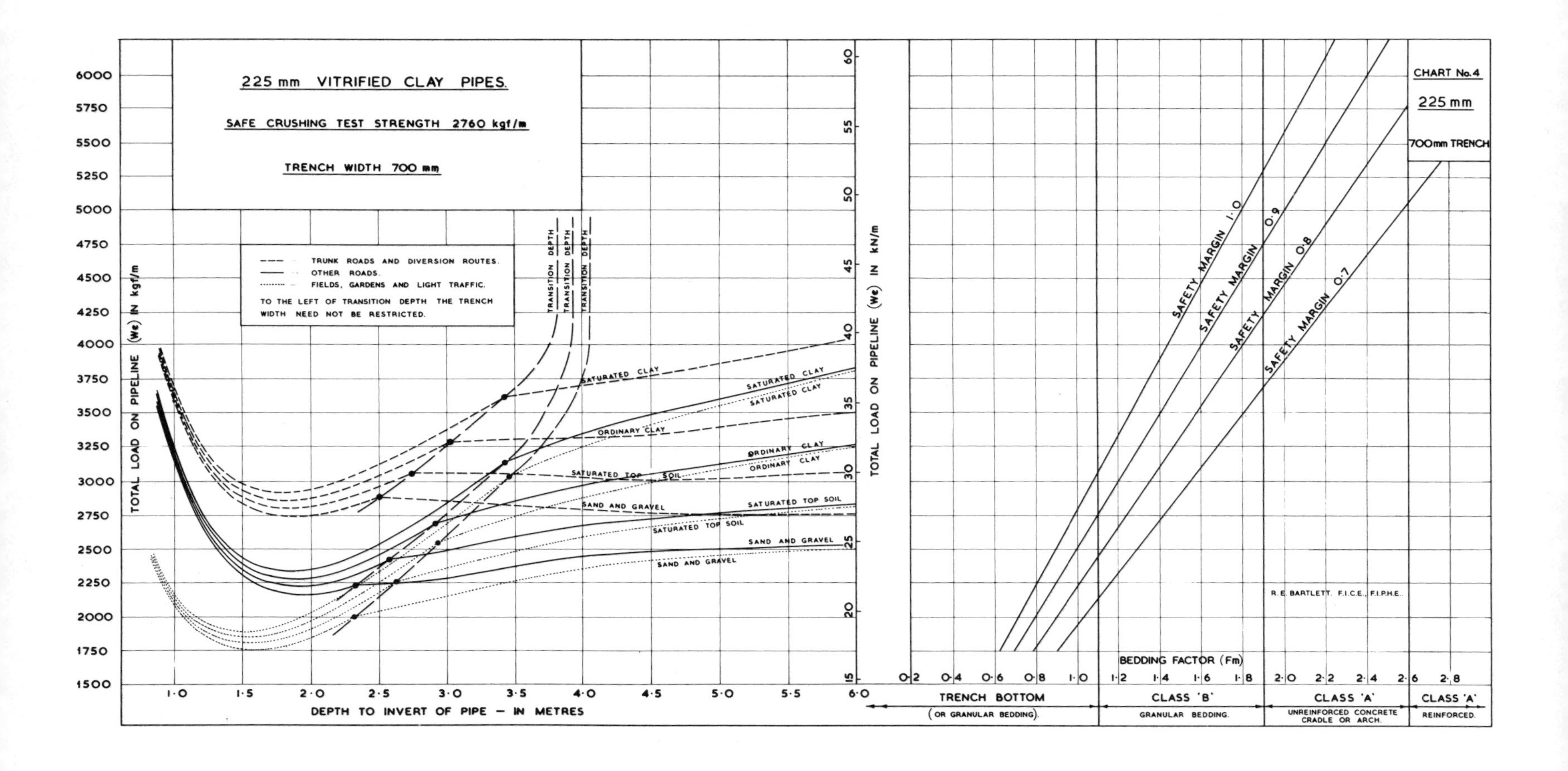
225 mm VITRIFIED CLAY PIPES.
SAFE CRUSHING TEST STRENGTH 2760 kgf/m
TRENCH WIDTH 700 mm
CHART No. 4
225 mm
700mm TRENCH
TRUNK ROADS AND DIVERSION ROUTES.
OTHER ROADS.
FIELDS, GARDENS AND LIGHT TRAFFIC.
TO THE LEFT OF TRANSITION DEPTH THE TRENCH WIDTH NEED NOT BE RESTRICTED.
TRANSITION DEPTH
TOTAL LOAD ON PIPELINE (We) IN kgf/m
TOTAL LOAD ON PIPELINE (We) IN kN/m
DEPTH TO INVERT OF PIPE — IN METRES
SATURATED CLAY
ORDINARY CLAY
SATURATED TOP SOIL
SAND AND GRAVEL
SAFETY MARGIN 1·0
SAFETY MARGIN 0·9
SAFETY MARGIN 0·8
SAFETY MARGIN 0·7
BEDDING FACTOR (Fm)
TRENCH BOTTOM
(OR GRANULAR BEDDING).
CLASS 'B'
GRANULAR BEDDING.
CLASS 'A'
UNREINFORCED CONCRETE CRADLE OR ARCH.
CLASS 'A'
REINFORCED.
R. E. BARTLETT. F.I.C.E., F.I.P.H.E.

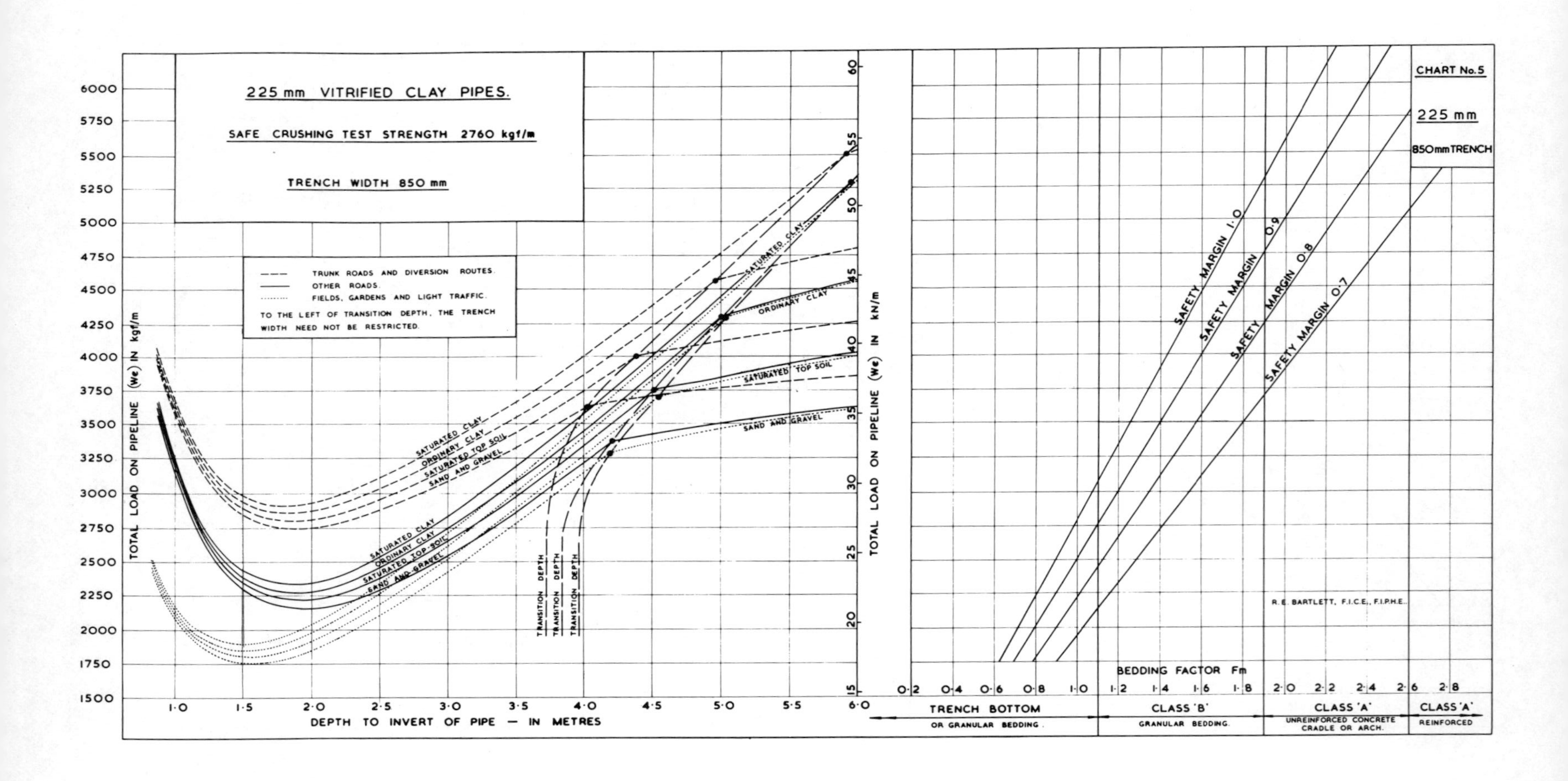

CHART No. 5
225 mm
850 mm TRENCH
225 mm VITRIFIED CLAY PIPES.
SAFE CRUSHING TEST STRENGTH 2760 kgf/m
TRENCH WIDTH 850 mm
TRUNK ROADS AND DIVERSION ROUTES.
OTHER ROADS.
FIELDS, GARDENS AND LIGHT TRAFFIC.
TO THE LEFT OF TRANSITION DEPTH, THE TRENCH WIDTH NEED NOT BE RESTRICTED.
TOTAL LOAD ON PIPELINE (We) IN kgf/m
6000
5750
5500
5250
5000
4750
4500
4250
4000
3750
3500
3250
3000
2750
2500
2250
2000
1750
1500
1·0
1·5
2·0
2·5
3·0
3·5
4·0
4·5
5·0
5·5
6·0
DEPTH TO INVERT OF PIPE — IN METRES
SATURATED CLAY
ORDINARY CLAY
SATURATED TOP SOIL
SAND AND GRAVEL
TRANSITION DEPTH
TRANSITION DEPTH
TRANSITION DEPTH
TOTAL LOAD ON PIPELINE (We) IN kN/m
15
20
25
30
35
40
45
50
55
60
SAFETY MARGIN 1·0
SAFETY MARGIN 0·9
SAFETY MARGIN 0·8
SAFETY MARGIN 0·7
R. E. BARTLETT, F.I.C.E., F.I.P.H.E.
BEDDING FACTOR Fm
0·2
0·4
0·6
0·8
1·0
1·2
1·4
1·6
1·8
2·0
2·2
2·4
2·6
2·8
TRENCH BOTTOM
OR GRANULAR BEDDING.
CLASS 'B'
GRANULAR BEDDING.
CLASS 'A'
UNREINFORCED CONCRETE CRADLE OR ARCH.
CLASS 'A'
REINFORCED

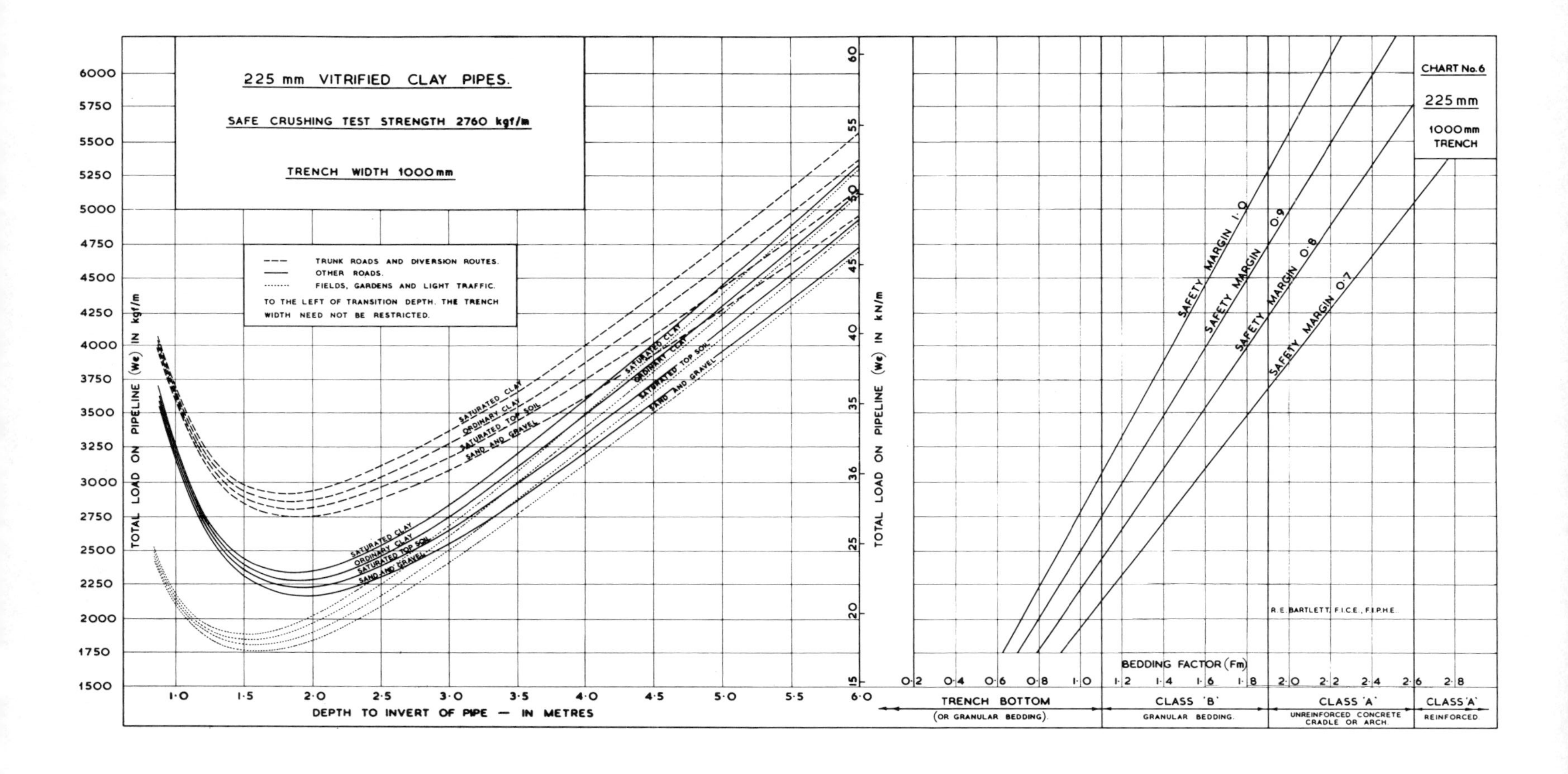

225 mm VITRIFIED CLAY PIPES.
SAFE CRUSHING TEST STRENGTH 2760 kgf/m
TRENCH WIDTH 1000 mm
TRUNK ROADS AND DIVERSION ROUTES.
OTHER ROADS.
FIELDS, GARDENS AND LIGHT TRAFFIC.
TO THE LEFT OF TRANSITION DEPTH. THE TRENCH WIDTH NEED NOT BE RESTRICTED.
TOTAL LOAD ON PIPELINE (We) IN kgf/m
6000
5750
5500
5250
5000
4750
4500
4250
4000
3750
3500
3250
3000
2750
2500
2250
2000
1750
1500
SATURATED CLAY
ORDINARY CLAY
SATURATED TOP SOIL
SAND AND GRAVEL
DEPTH TO INVERT OF PIPE — IN METRES
1·0
1·5
2·0
2·5
3·0
3·5
4·0
4·5
5·0
5·5
6·0
TOTAL LOAD ON PIPELINE (We) IN kN/m
15
20
25
36
35
40
45
50
55
60
SAFETY MARGIN 1·0
SAFETY MARGIN 0·9
SAFETY MARGIN 0·8
SAFETY MARGIN 0·7
CHART No. 6
225 mm
1000 mm TRENCH
R.E. BARTLETT, F.I.C.E., F.I.P.H.E.
BEDDING FACTOR (Fm)
0·2
0·4
0·6
0·8
1·0
1·2
1·4
1·6
1·8
2·0
2·2
2·4
2·6
2·8
TRENCH BOTTOM
(OR GRANULAR BEDDING).
CLASS 'B'
GRANULAR BEDDING.
CLASS 'A'
UNREINFORCED CONCRETE CRADLE OR ARCH.
CLASS 'A'
REINFORCED.

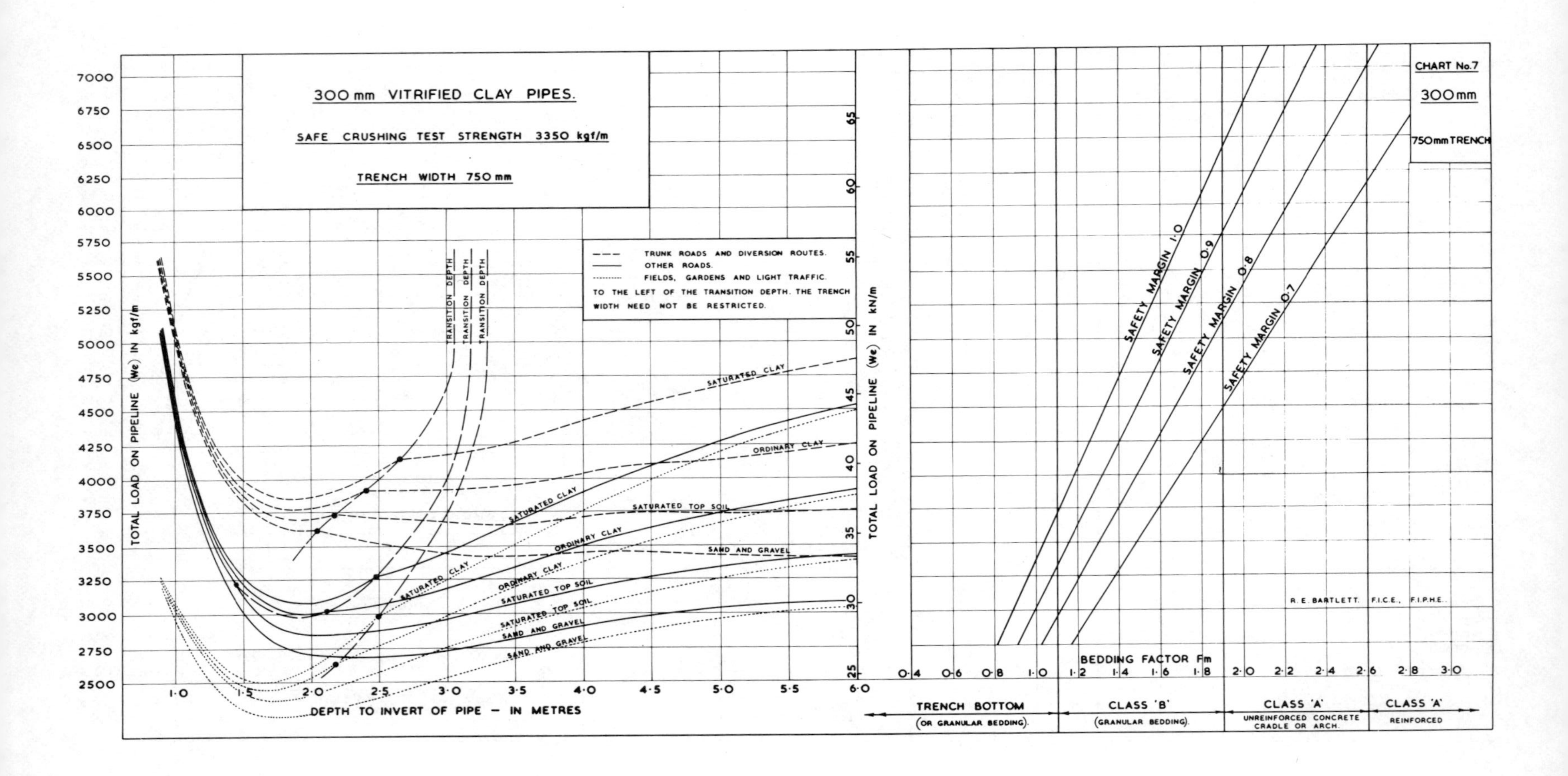

300 mm VITRIFIED CLAY PIPES.
SAFE CRUSHING TEST STRENGTH 3350 kgf/m
TRENCH WIDTH 750 mm
CHART No.7
300 mm
750 mm TRENCH
TRUNK ROADS AND DIVERSION ROUTES.
OTHER ROADS.
FIELDS, GARDENS AND LIGHT TRAFFIC.
TO THE LEFT OF THE TRANSITION DEPTH. THE TRENCH WIDTH NEED NOT BE RESTRICTED.
TRANSITION DEPTH
TRANSITION DEPTH
TRANSITION DEPTH
SATURATED CLAY
ORDINARY CLAY
SATURATED TOP SOIL
SAND AND GRAVEL
SATURATED CLAY
ORDINARY CLAY
SATURATED CLAY
ORDINARY CLAY
SATURATED TOP SOIL
SATURATED TOP SOIL
SAND AND GRAVEL
SAND AND GRAVEL
TOTAL LOAD ON PIPELINE (We) IN kgf/m
7000
6750
6500
6250
6000
5750
5500
5250
5000
4750
4500
4250
4000
3750
3500
3250
3000
2750
2500
1·0
1·5
2·0
2·5
3·0
3·5
4·0
4·5
5·0
5·5
6·0
DEPTH TO INVERT OF PIPE – IN METRES
TOTAL LOAD ON PIPELINE (We) IN kN/m
65
60
55
50
45
40
35
30
25
SAFETY MARGIN 1·0
SAFETY MARGIN 0·9
SAFETY MARGIN 0·8
SAFETY MARGIN 0·7
R. E. BARTLETT. F.I.C.E., F.I.P.H.E.
BEDDING FACTOR Fm
0·4
0·6
0·8
1·0
1·2
1·4
1·6
1·8
2·0
2·2
2·4
2·6
2·8
3·0
TRENCH BOTTOM
(OR GRANULAR BEDDING).
CLASS 'B'
(GRANULAR BEDDING).
CLASS 'A'
UNREINFORCED CONCRETE CRADLE OR ARCH.
CLASS 'A'
REINFORCED

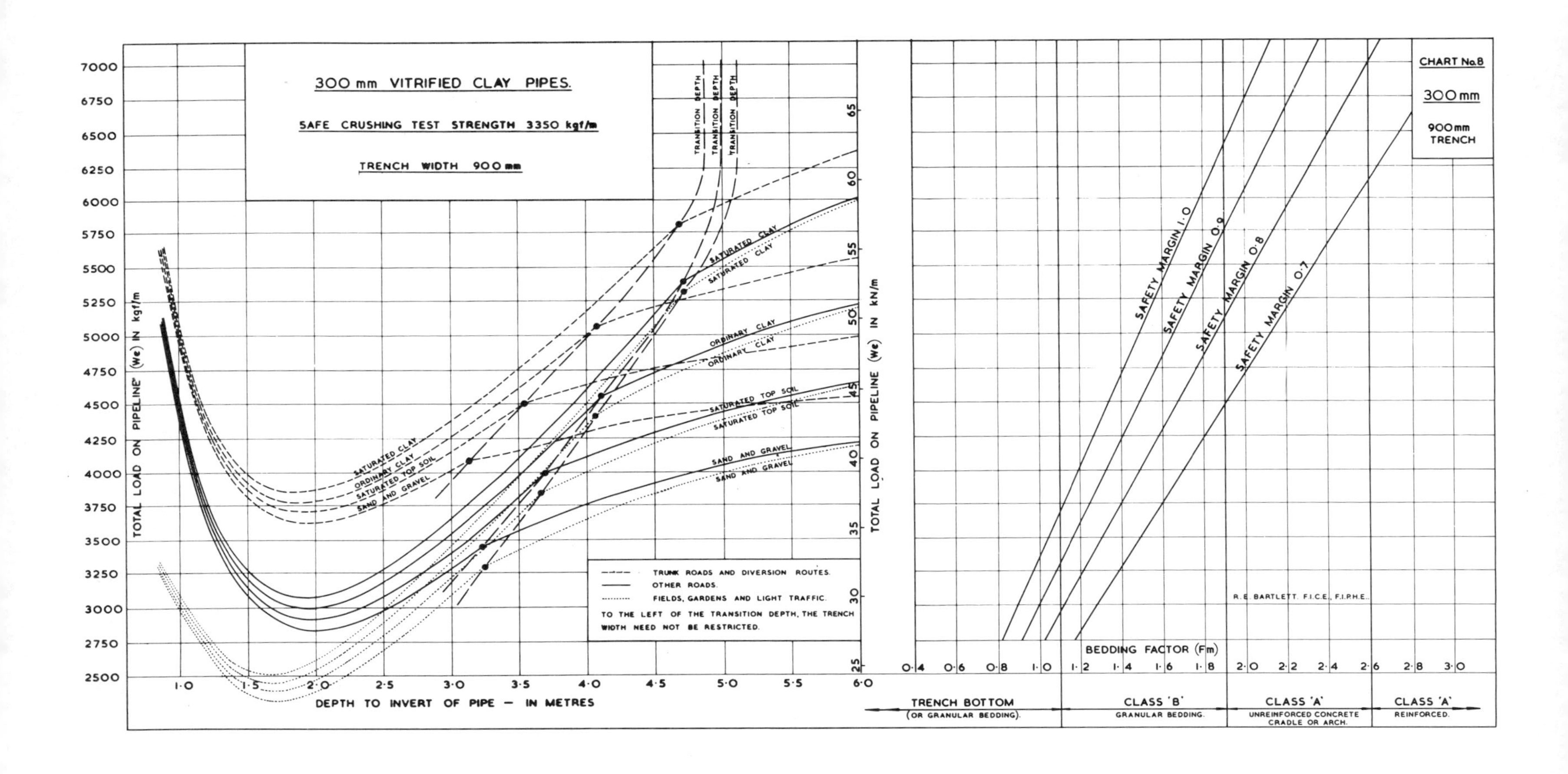
CHART No.8
300 mm
900 mm TRENCH
300 mm VITRIFIED CLAY PIPES.
SAFE CRUSHING TEST STRENGTH 3350 kgf/m
TRENCH WIDTH 900 mm
TRANSITION DEPTH
TRANSITION DEPTH
TRANSITION DEPTH
SATURATED CLAY
ORDINARY CLAY
SATURATED TOP SOIL
SAND AND GRAVEL
SATURATED CLAY
SATURATED CLAY
ORDINARY CLAY
ORDINARY CLAY
SATURATED TOP SOIL
SATURATED TOP SOIL
SAND AND GRAVEL
SAND AND GRAVEL
TRUNK ROADS AND DIVERSION ROUTES.
OTHER ROADS.
FIELDS, GARDENS AND LIGHT TRAFFIC.
TO THE LEFT OF THE TRANSITION DEPTH, THE TRENCH WIDTH NEED NOT BE RESTRICTED.
TOTAL LOAD ON PIPELINE (We) IN kgf/m
7000
6750
6500
6250
6000
5750
5500
5250
5000
4750
4500
4250
4000
3750
3500
3250
3000
2750
2500
1·0
1·5
2·0
2·5
3·0
3·5
4·0
4·5
5·0
5·5
6·0
DEPTH TO INVERT OF PIPE — IN METRES
TOTAL LOAD ON PIPELINE (We) IN kN/m
65
60
55
50
45
40
35
30
25
SAFETY MARGIN 1·0
SAFETY MARGIN 0·9
SAFETY MARGIN 0·8
SAFETY MARGIN 0·7
R. E. BARTLETT. F.I.C.E., F.I.P.H.E.
BEDDING FACTOR (Fm)
0·4
0·6
0·8
1·0
1·2
1·4
1·6
1·8
2·0
2·2
2·4
2·6
2·8
3·0
TRENCH BOTTOM
(OR GRANULAR BEDDING).
CLASS 'B'
GRANULAR BEDDING.
CLASS 'A'
UNREINFORCED CONCRETE CRADLE OR ARCH.
CLASS 'A'
REINFORCED.

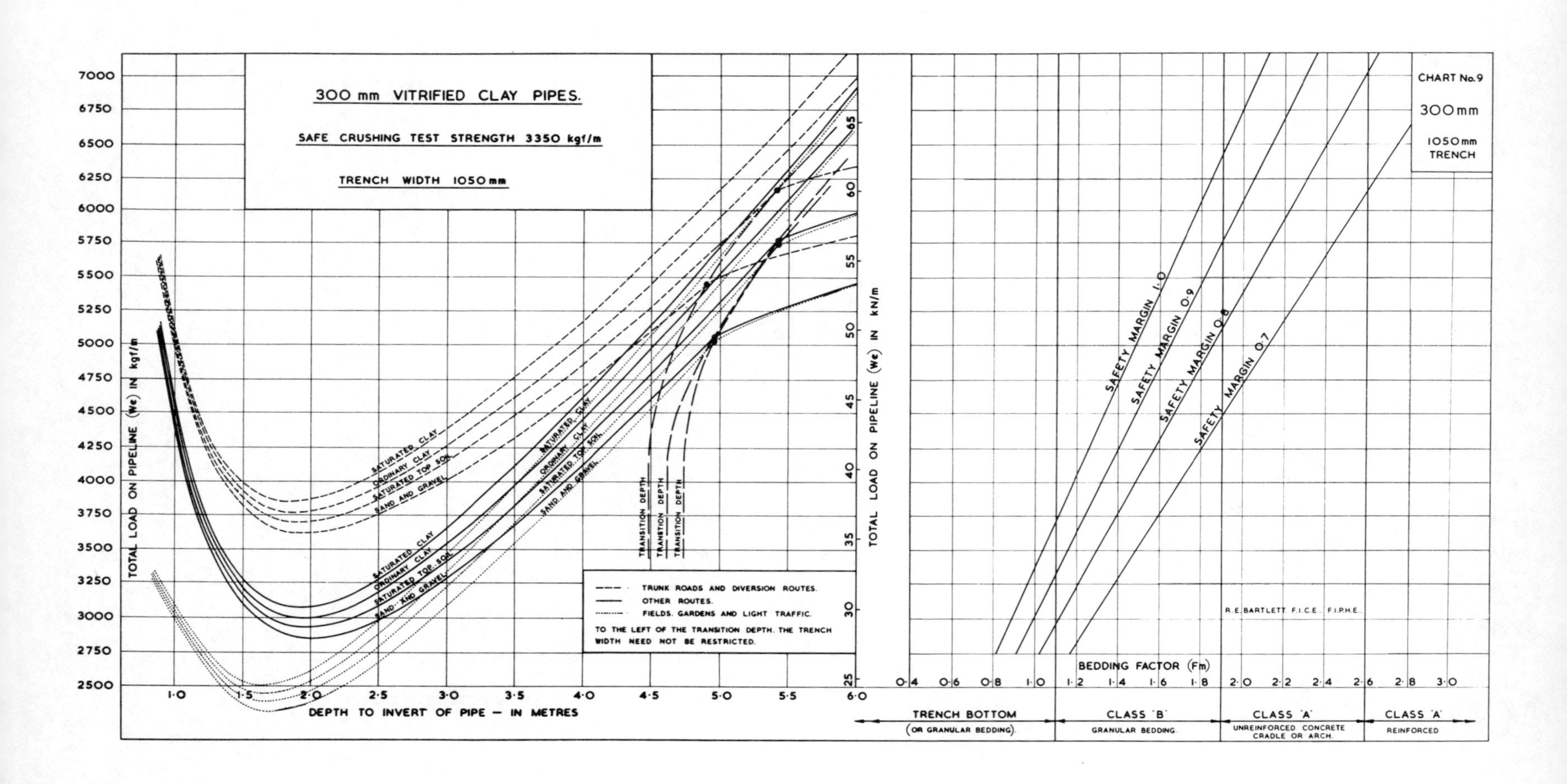
300 mm VITRIFIED CLAY PIPES.
SAFE CRUSHING TEST STRENGTH 3350 kgf/m
TRENCH WIDTH 1050 mm
CHART No. 9
300 mm
1050 mm TRENCH
TOTAL LOAD ON PIPELINE (We) IN kgf/m
7000
6750
6500
6250
6000
5750
5500
5250
5000
4750
4500
4250
4000
3750
3500
3250
3000
2750
2500
DEPTH TO INVERT OF PIPE — IN METRES
1·0
1·5
2·0
2·5
3·0
3·5
4·0
4·5
5·0
5·5
6·0
TOTAL LOAD ON PIPELINE (We) IN kN/m
25
30
35
40
45
50
55
60
65
SATURATED CLAY
ORDINARY CLAY
SATURATED TOP SOIL
SAND AND GRAVEL
TRANSITION DEPTH
TRANSITION DEPTH
TRANSITION DEPTH
TRUNK ROADS AND DIVERSION ROUTES.
OTHER ROUTES.
FIELDS, GARDENS AND LIGHT TRAFFIC.
TO THE LEFT OF THE TRANSITION DEPTH, THE TRENCH WIDTH NEED NOT BE RESTRICTED.
SAFETY MARGIN 1·0
SAFETY MARGIN 0·9
SAFETY MARGIN 0·8
SAFETY MARGIN 0·7
R. E. BARTLETT F.I.C.E., F.I.P.H.E.
BEDDING FACTOR (Fm)
0·4
0·6
0·8
1·0
1·2
1·4
1·6
1·8
2·0
2·2
2·4
2·6
2·8
3·0
TRENCH BOTTOM
(OR GRANULAR BEDDING).
CLASS 'B'
GRANULAR BEDDING.
CLASS 'A'
UNREINFORCED CONCRETE CRADLE OR ARCH.
CLASS 'A'
REINFORCED

Index